Origin of l

Compiled by

Hugh Coon

Scribbles

Year of Publication 2018

ISBN : 9789352979264

Book Published by

Scribbles

(An Imprint of Alpha Editions)

email - alphaedis@gmail.com

Produced by: PediaPress GmbH
Limburg an der Lahn
Germany
http://pediapress.com/

Contents

Introduction

Abiogenesis

Abiogenesis, or informally the **origin of life**,[1,2] is the natural process by which life arises from non-living matter, such as simple organic compounds. The transition from non-living to living entities was not a single event, but a gradual process of increasing complexity.

Researchers study abiogenesis through a combination of molecular biology, paleontology, astrobiology and biochemistry, and aim to determine how pre-life chemical reactions gave rise to life. The study of abiogenesis can be geo-physical, chemical, or biological, with more recent approaches attempting a synthesis of all three, as life arose under conditions that are strikingly different from those on Earth today. Life functions through the specialized chemistry of carbon and water and builds largely upon four key families of chemicals: lipids (fatty cell walls), carbohydrates (sugars, cellulose), amino acids (protein metabolism), and nucleic acids (self-replicating DNA and RNA). Any successful theory of abiogenesis must explain the origins and interactions of these classes of molecules. Many approaches to abiogenesis investigate how self-replicating molecules, or their components, came into existence. Researchers generally think that current life on Earth descends from an RNA world, although RNA-based life may not have been the first life to have existed.

The classic 1952 Miller–Urey experiment and similar research demonstrated that most amino acids, the chemical constituents of the proteins used in all living organisms, can be synthesized from inorganic compounds under conditions intended to replicate those of the early Earth. Scientists have proposed various external sources of energy that may have triggered these reactions, including lightning and radiation. Other approaches ("metabolism-first" hypotheses) focus on understanding how catalysis in chemical systems on the

1

Figure 1: *The earliest known life-forms on Earth are putative fossilized microorganisms, found in hydrothermal vent precipitates, that may have lived as early as 4.28 billion years ago, relatively soon after the oceans formed 4.41 billion years ago, and not long after the formation of the Earth 4.54 billion years ago.*

early Earth might have provided the precursor molecules necessary for self-replication. Complex organic molecules occur in the Solar System and in interstellar space, and these molecules may have provided starting material for the development of life on Earth.

The biochemistry of life may have begun shortly after the Big Bang, 13.8 billion years ago, during a habitable epoch when the age of the universe was only 10 to 17 million years. The panspermia hypothesis suggests that microscopic life was distributed to the early Earth by space dust, meteoroids, asteroids and other small Solar System bodies and that life may exist throughout the universe.[3] The panspermia hypothesis proposes that life originated outside the Earth, but does not definitively explain its origin.

Nonetheless, Earth remains the only place in the universe known to harbour life, and fossil evidence from the Earth informs most studies of abiogenesis. The age of the Earth is about 4.54 billion years; the earliest undisputed evidence of life on Earth dates from at least 3.5 billion years ago, and possibly as early as the Eoarchean Era (between 3.6 and 4.0 billion years ago), after geological crust started to solidify following the molten Hadean Eon. In May 2017 scientists found possible evidence of early life on land in 3.48-billion-year-old geyserite and other related mineral deposits (often found around hot springs and geysers) uncovered in the Pilbara Craton of Western Australia. However, a number of discoveries suggest that life may have appeared on Earth even earlier. As of 2017[4], microfossils within hydrothermal-vent precipitates dated from 3.77 to 4.28 billion years old found in Quebec, Canada may be the oldest record of life on Earth, suggesting life started soon after ocean formation 4.4 billion years ago. According to biologist Stephen Blair Hedges, "If life arose relatively quickly on Earth ... then it could be common in the universe."

Early geophysical conditions on Earth

Life timeline

0 —
500 —
1000 —
1500 —
2000 —
2500 —
3000 —
3500 —
4000 —
4500 —

Axis scale: million years

🖐

Also see: *Human timeline* and *Nature timeline*

The Hadean Earth is thought to have had a secondary atmosphere, formed through degassing of the rocks that accumulated from planetesimal impactors. At first, it was thought that the Earth's atmosphere consisted of hydrogen compounds—methane, ammonia and water vapour—and that life began under such reducing conditions, which are conducive to the formation of organic molecules. According to later models, suggested by study of ancient minerals, the atmosphere in the late Hadean period consisted largely of water vapour, nitrogen and carbon dioxide, with smaller amounts of carbon monoxide, hydrogen, and sulfur compounds. During its formation, the Earth lost a significant part of its initial mass, with a nucleus of the heavier rocky elements of the protoplanetary disk remaining. As a consequence, Earth lacked the gravity to hold any molecular hydrogen in its atmosphere, and rapidly lost it during the Hadean period, along with the bulk of the original inert gases. The solution of carbon dioxide in water is thought to have made the seas slightly acidic, giving it a pH of about 5.5.Wikipedia:Citation needed The atmosphere at the time has been characterized as a "gigantic, productive outdoor chemical laboratory." It may have been similar to the mixture of gases released today by volcanoes, which still support some abiotic chemistry.

Oceans may have appeared first in the Hadean Eon, as soon as two hundred million years (200 Ma) after the Earth was formed, in a hot 100 °C (212 °F) reducing environment, and the pH of about 5.8 rose rapidly towards neutral. This has been supported by the dating of 4.404 Ga-old zircon crystals from metamorphosed quartzite of Mount Narryer in the Western Australia Jack Hills of the Pilbara, which are evidence that oceans and continental crust existed within 150 Ma of Earth's formation. Despite the likely increased volcanism and existence of many smaller tectonic "platelets," it has been suggested that between 4.4 and 4.3 Ga (billion year), the Earth was a water world, with little if any continental crust, an extremely turbulent atmosphere and a hydrosphere subject to intense ultraviolet (UV) light, from a T Tauri stage Sun, cosmic radiation and continued bolide impacts.

The Hadean environment would have been highly hazardous to modern life. Frequent collisions with large objects, up to 500 kilometres (310 mi) in diameter, would have been sufficient to sterilize the planet and vaporize the ocean within a few months of impact, with hot steam mixed with rock vapour becoming high altitude clouds that would completely cover the planet. After a few months, the height of these clouds would have begun to decrease but the cloud base would still have been elevated for about the next thousand years. After that, it would have begun to rain at low altitude. For another two thousand years, rains would slowly have drawn down the height of the clouds, returning the oceans to their original depth only 3,000 years after the impact event.

Earliest biological evidence for life

The most commonly accepted location of the root of the tree of life is between a monophyletic domain Bacteria and a clade formed by Archaea and Eukaryota of what is referred to as the "traditional tree of life" based on several molecular studies starting with C. Woese. A very small minority of studies have concluded differently, namely that the root is in the Domain Bacteria, either in the phylum Firmicutes or that the phylum Chloroflexi is basal to a clade with Archaea+Eukaryotes and the rest of Bacteria as proposed by Thomas Cavalier-Smith. More recently Peter Ward has established an alternative view which is rooted in abiotic RNA synthesis which becomes enclosed within a capsule and then creates RNA ribozyme replicates. It is proposed that this then bifurcates between Dominion Ribosa (RNA life), and after the loss of ribozymes RNA viruses as Domain Viorea, and Dominion TerroaWikipedia:Please clarify, which after creating a large cell within a lipid wall, creating DNA the 20 based amino acids and the triplet code, is established as the last universal common ancestor or LUCA, of earlier phylogenic trees.[5]

The earliest life on Earth existed more than 3.5 billion years ago, during the Eoarchean Era when sufficient crust had solidified following the molten Hadean Eon. The earliest physical evidence so far found consists of microfossils in the Nuvvuagittuq Greenstone Belt of Northern Quebec, in "banded iron formation" rocks at least 3.77 billion and possibly 4.28 billion years old. This finding suggested that there was almost instant development of life after oceans were formed. The structure of the microbes was noted to be similar to bacteria found near hydrothermal vents in the modern era, and provided support for the hypothesis that abiogenesis began near hydrothermal vents.

Also noteworthy is biogenic graphite in 3.7 billion-year-old metasedimentary rocks from southwestern Greenland and microbial mat fossils found in 3.48 billion-year-old sandstone from Western Australia. Evidence of early life in rocks from Akilia Island, near the Isua supracrustal belt in southwestern Greenland, dating to 3.7 billion years ago have shown biogenic carbon isotopes. In other parts of the Isua supracrustal belt, graphite inclusions trapped within garnet crystals are connected to the other elements of life: oxygen, nitrogen, and possibly phosphorus in the form of phosphate, providing further evidence for life 3.7 billion years ago. At Strelley Pool, in the Pilbara region of Western Australia, compelling evidence of early life was found in pyrite-bearing sandstone in a fossilized beach, that showed rounded tubular cells that oxidized sulfur by photosynthesis in the absence of oxygen. Further research on zircons

Figure 2: *Precambrian stromatolites in the Siyeh Formation, Glacier National Park. In 2002, a paper in the scientific journal Nature suggested that these 3.5 Ga (billion years) old geological formations contain fossilized cyanobacteria microbes. This suggests they are evidence of one of the earliest life forms on Earth.*

from Western Australia in 2015 suggested that life likely existed on Earth at least 4.1 billion years ago.[6]

Traditionally it was thought that during the period between 4.28 and 3.8 Ga, changes in the orbits of the giant planets may have caused a heavy bombardment by asteroids and comets that pockmarked the Moon and the other inner planets (Mercury, Mars, and presumably Earth and Venus). This would likely have repeatedly sterilized the planet, had life appeared before that time. Geologically, the Hadean Earth would have been far more active than at any other time in its history. Studies of meteorites suggests that radioactive isotopes such as aluminium-26 with a half-life of 7.17×10^5 (717 thousand) years, and potassium-40 with a half-life of 1.250×10^9 (1.25 billion) years, isotopes mainly produced in supernovae, were much more common. Internal heating as a result of gravitational sorting between the core and the mantle would have caused a great deal of mantle convection, with the probable result of many more smaller and more active tectonic plates than now exist.

The time periods between such devastating environmental events give time windows for the possible origin of life in the early environments. If the deep marine hydrothermal setting was the site for the origin of life, then abiogenesis could have happened as early as 4.0 to 4.2 Ga. If the site was at the surface of the Earth, abiogenesis could only have occurred between 3.7 and 4.0 Ga.

In 2016, a set of 355 genes likely present in the Last Universal Common Ancestor (LUCA) of all organisms living on Earth was identified. A total of 6.1 million prokaryotic protein coding genes from various phylogenic trees were sequenced, identifying 355 protein clusters from amongst 286,514 protein clusters that were probably common to LUCA. The results "depict LUCA as anaerobic, CO_2-fixing, H_2-dependent with a Wood–Ljungdahl pathway, N_2-fixing and thermophilic. LUCA's biochemistry was replete with FeS clusters and radical reaction mechanisms. Its cofactors reveal dependence upon transition metals, flavins, S-adenosyl methionine, coenzyme A, ferredoxin, molybdopterin, corrins and selenium. Its genetic code required nucleoside modifications and S-adenosylmethionine-dependent methylations." The results depict methanogenic clostridia as a basal clade in the 355 phylogenies examined, and suggest that LUCA inhabited an anaerobic hydrothermal vent setting in a geochemically active environment rich in H_2, CO_2 and iron. M.D. Brazier has shown that the tiny fossils discovered came from a hot poisonous world of the toxic gases methane, ammonia, carbon dioxide and hydrogen sulphide.[7] An analysis of the conventional threefold tree of life shows thermophilic and hyperthermophilic bacteria and archaea are closest to the root, suggesting that life may have evolved in a hot environment.[8]

Conceptual history

Spontaneous generation

Belief in spontaneous generation of certain forms of life from non-living matter goes back to Aristotle and ancient Greek philosophy and continued to have support in Western scholarship until the 19th century. This belief was paired with a belief in heterogenesis, i.e., that one form of life derived from a different form (e.g., bees from flowers). Classical notions of spontaneous generation held that certain complex, living organisms are generated by decaying organic substances. According to Aristotle, it was a readily observable truth that aphids arise from the dew that falls on plants, flies from putrid matter, mice from dirty hay, crocodiles from rotting logs at the bottom of bodies of water, and so on. In the 17th century, people began to question such assumptions. In 1646, Sir Thomas Browne published his *Pseudodoxia Epidemica* (subtitled *Enquiries into Very many Received Tenets, and commonly Presumed Truths*), which was an attack on false beliefs and "vulgar errors." His contemporary, Alexander Ross, erroneously refuted him, stating: "To question this [spontaneous generation], is to question Reason, Sense, and Experience: If he doubts of this, let him go to *Ægypt*, and there he will finde the fields swarming with mice begot of the mud of *Nylus*, to the great calamity of the Inhabitants." In 1665, Robert Hooke published the first drawings of a microorganism. Hooke was followed in 1676 by Antonie van Leeuwenhoek, who drew and described microorganisms that are now thought to have been protozoa and bacteria. Many felt the existence of microorganisms was evidence in support of spontaneous generation, since microorganisms seemed too simplistic for sexual reproduction, and asexual reproduction through cell division had not yet been observed. Van Leeuwenhoek took issue with the ideas common at the time that fleas and lice could spontaneously result from putrefaction, and that frogs could likewise arise from slime. Using a broad range of experiments ranging from sealed and open meat incubation and the close study of insect reproduction he became, by the 1680s, convinced that spontaneous generation was incorrect.

The first experimental evidence against spontaneous generation came in 1668 when Francesco Redi showed that no maggots appeared in meat when flies were prevented from laying eggs. It was gradually shown that, at least in the case of all the higher and readily visible organisms, the previous sentiment regarding spontaneous generation was false. The alternative seemed to be biogenesis: that every living thing came from a pre-existing living thing (*omne vivum ex ovo*, Latin for "every living thing from an egg").

In 1768, Lazzaro Spallanzani demonstrated that microbes were present in the air, and could be killed by boiling. In 1861, Louis Pasteur performed a series of experiments that demonstrated that organisms such as bacteria and fungi do not spontaneously appear in sterile, nutrient-rich media, but could only appear by invasion from without.

The belief that self-ordering by spontaneous generation was impossible begged for an alternative. By the middle of the 19th century, the theory of biogenesis had accumulated so much evidential support, due to the work of Pasteur and others, that the alternative theory of spontaneous generation had been effectively disproven. John Desmond Bernal, a pioneer in X-ray crystallography, suggested that earlier theories such as spontaneous generation were based upon an explanation that life was continuously created as a result of chance events.

Etymology

The term *biogenesis* is usually credited to either Henry Charlton Bastian or to Thomas Henry Huxley. Bastian used the term around 1869 in an unpublished exchange with John Tyndall to mean "life-origination or commencement". In 1870, Huxley, as new president of the British Association for the Advancement of Science, delivered an address entitled *Biogenesis and Abiogenesis*. In it he introduced the term *biogenesis* (with an opposite meaning to Bastian's) as well as *abiogenesis*:

> And thus the hypothesis that living matter always arises by the agency of pre-existing living matter, took definite shape; and had, henceforward, a right to be considered and a claim to be refuted, in each particular case, before the production of living matter in any other way could be admitted by careful reasoners. It will be necessary for me to refer to this hypothesis so frequently, that, to save circumlocution, I shall call it the hypothesis of *Biogenesis*; and I shall term the contrary doctrine–that living matter may be produced by not living matter–the hypothesis of *Abiogenesis*.

Subsequently, in the preface to Bastian's 1871 book, *The Modes of Origin of Lowest Organisms*, Bastian referred to the possible confusion with Huxley's usage and explicitly renounced his own meaning:

> A word of explanation seems necessary with regard to the introduction of the new term *Archebiosis*. I had originally, in unpublished writings, adopted the word *Biogenesis* to express the same meaning—viz., life-origination or commencement. But in the mean time the word *Biogenesis* has been made use of, quite independently, by a distinguished biologist [Huxley], who wished to make it bear a totally different meaning. He also introduced the word *Abiogenesis*. I have been informed, however, on the best authority, that neither of these words can—with any regard to the language from which they are derived—be supposed to bear the meanings which have of late been publicly assigned to them. Wishing to avoid all needless confusion, I therefore renounced the use of the word *Biogenesis*, and being, for the reason just given, unable to adopt the other term, I was compelled to introduce a new word, in order to designate the process by which living matter is supposed to come into being, independently of pre-existing living matter.

Louis Pasteur and Charles Darwin

Louis Pasteur remarked, about a finding of his in 1864 which he considered definitive, "Never will the doctrine of spontaneous generation recover from the mortal blow struck by this simple experiment." One alternative was that life's origins on Earth had come from somewhere else in the universe. Periodically resurrected (see Panspermia, above) Bernal said that this approach "is equivalent in the last resort to asserting the operation of metaphysical, spiritual entities... it turns on the argument of creation by design by a creator or demiurge." Such a theory, Bernal said, was unscientific. A theory popular around the same time was that life was the result of an inner "life force", which in the late 19th century was championed by Henri Bergson.

Figure 3: *Charles Darwin in 1879*

The idea of evolution by natural selection proposed by Charles Darwin put
an end to these metaphysical theologies. In a letter to Joseph Dalton Hooker
on 1 February 1871, Darwin discussed the suggestion that the original spark
of life may have begun in a "warm little pond, with all sorts of ammonia and
phosphoric salts, light, heat, electricity, &c., present, that a proteine compound
was chemically formed ready to undergo still more complex changes." He went
on to explain that "at the present day such matter would be instantly devoured
or absorbed, which would not have been the case before living creatures were
formed." He had written to Hooker in 1863 stating that, "It is mere rubbish,
thinking at present of the origin of life; one might as well think of the origin
of matter." In *On the Origin of Species*, he had referred to life having been
"created", by which he "really meant 'appeared' by some wholly unknown
process", but had soon regretted using the Old Testament term "creation".[9]

"Primordial soup" hypothesis

No new notable research or hypothesis on the subject appeared until 1924,
when Alexander Oparin reasoned that atmospheric oxygen prevents the synthesis of certain organic compounds that are necessary building blocks for
the evolution of life. In his book *The Origin of Life*, Oparin proposed that
the "spontaneous generation of life" that had been attacked by Louis Pasteur
did in fact occur once, but was now impossible because the conditions found

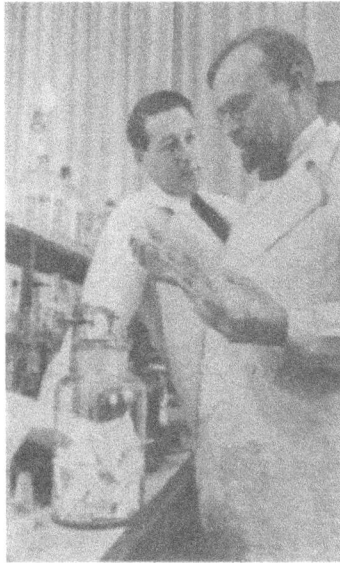

Figure 4: *Alexander Oparin (right) in his laboratory, 1938*

on the early Earth had changed, and preexisting organisms would immediately consume any spontaneously generated organism. Oparin argued that a "primeval soup" of organic molecules could be created in an oxygenless atmosphere through the action of sunlight. These would combine in ever more complex ways until they formed coacervate droplets. These droplets would "grow" by fusion with other droplets, and "reproduce" through fission into daughter droplets, and so have a primitive metabolism in which factors that promote "cell integrity" survive, and those that do not become extinct. Many modern theories of the origin of life still take Oparin's ideas as a starting point.

Robert Shapiro has summarized the "primordial soup" theory of Oparin and J. B. S. Haldane in its "mature form" as follows:

1. The early Earth had a chemically reducing atmosphere.
2. This atmosphere, exposed to energy in various forms, produced simple organic compounds ("monomers").
3. These compounds accumulated in a "soup" that may have concentrated at various locations (shorelines, oceanic vents etc.).
4. By further transformation, more complex organic polymers – and ultimately life – developed in the soup.

About this time, Haldane suggested that the Earth's prebiotic oceans (quite different from their modern counterparts) would have formed a "hot dilute soup" in which organic compounds could have formed. Bernal called this idea *biopoiesis* or *biopoesis*, the process of living matter evolving from self-replicating but non-living molecules, and proposed that biopoiesis passes through a number of intermediate stages.

One of the most important pieces of experimental support for the "soup" theory came in 1952. Stanley L. Miller and Harold C. Urey performed an experiment that demonstrated how organic molecules could have spontaneously formed from inorganic precursors under conditions like those posited by the Oparin-Haldane hypothesis. The now-famous Miller–Urey experiment used a highly reducing mixture of gases – methane, ammonia, and hydrogen, as well as water vapour – to form simple organic monomers such as amino acids. The mixture of gases was cycled through an apparatus that delivered electrical sparks to the mixture. After one week, it was found that about 10% to 15% of the carbon in the system was then in the form of a racemic mixture of organic compounds, including amino acids, which are the building blocks of proteins. This provided direct experimental support for the second point of the "soup" theory, and it is around the remaining two points of the theory that much of the debate now centres.

Bernal showed that based upon this and subsequent work there is no difficulty in principle in forming most of the molecules we recognize as the necessary molecules for life from their inorganic precursors. The underlying hypothesis held by Oparin, Haldane, Bernal, Miller and Urey, for instance, was that multiple conditions on the primeval Earth favoured chemical reactions that synthesized the same set of complex organic compounds from such simple precursors. A 2011 reanalysis of the saved vials containing the original extracts that resulted from the Miller and Urey experiments, using current and more advanced analytical equipment and technology, has uncovered more biochemicals than originally discovered in the 1950s. One of the more important findings was 23 amino acids, far more than the five originally found. However, Bernal said that "it is not enough to explain the formation of such molecules, what is necessary, is a physical-chemical explanation of the origins of these molecules that suggests the presence of suitable sources and sinks for free energy."

More recent studies, in October 2017, support the notion that life may have begun right after the Earth was formed as RNA molecules emerging from "warm little ponds".

Proteinoid microspheres

In trying to uncover the intermediate stages of abiogenesis mentioned by Bernal, Sidney W. Fox in the 1950s and 1960s studied the spontaneous formation of peptide structures (small chains of amino acids) under conditions that might plausibly have existed early in Earth's history. In one of his experiments, he allowed amino acids to dry out as if puddled in a warm, dry spot in prebiotic conditions. He found that, as they dried, the amino acids formed long, often cross-linked, thread-like, submicroscopic polypeptide molecules now named "proteinoid microspheres".

In another experiment to set suitable conditions for life to form, Fox collected volcanic material from a cinder cone in Hawaii. He discovered that the temperature was over 100 °C (212 °F) just 4 inches (100 mm) beneath the surface of the cinder cone, and suggested that this might have been the environment in which life was created—molecules could have formed and then been washed through the loose volcanic ash into the sea. He placed lumps of lava over amino acids derived from methane, ammonia and water, sterilized all materials, and baked the lava over the amino acids for a few hours in a glass oven. A brown, sticky substance formed over the surface, and when the lava was drenched in sterilized water, a thick, brown liquid leached out. The amino acids had combined to form proteinoids, and the proteinoids had combined to form small globules that Fox called "microspheres". His proteinoids were not cells, although they formed clumps and chains reminiscent of cyanobacteria, but they contained no functional nucleic acids or any encoded information. Based upon such experiments, Colin S. Pittendrigh stated in December 1967 that "laboratories will be creating a living cell within ten years," a remark that reflected the typical contemporary naivety about the complexity of cell structures.

Current models

There is no single, generally accepted model for the origin of life. Scientists have proposed several plausible hypotheses, which share some common elements. While differing in the details, these hypotheses are based on the framework laid out by Alexander Oparin (in 1924) and by J. B. S. Haldane (in 1925), who postulated the molecular or chemical evolution theory of life. According to them, the first molecules constituting the earliest cells "were synthesized under natural conditions by a slow process of molecular evolution, and these molecules then organized into the first molecular system with properties with biological order". Oparin and Haldane suggested that the atmosphere of the early Earth may have been chemically reducing in nature, composed primarily of methane (CH_4), ammonia (NH_3), water (H_2O), hydrogen sulfide (H_2S), carbon dioxide (CO_2) or carbon monoxide (CO), and phosphate (PO_4^{3-}), with molecular oxygen (O_2) and ozone (O_3) either rare or absent. According to later models, the atmosphere in the late Hadean period consisted largely of nitrogen (N_2) and carbon dioxide, with smaller amounts of carbon monoxide, hydrogen (H_2), and sulfur compounds; while it did lack molecular oxygen and ozone, it was not as chemically reducing as Oparin and Haldane supposed. In the atmosphere proposed by Oparin and Haldane, electrical activity can produce certain small molecules (monomers) of life, such as amino acids. The Miller–Urey experiment reported in 1953 demonstrated this.

Bernal coined the term *biopoiesis* in 1949 to refer to the origin of life. In 1967, he suggested that it occurred in three "stages":

1. the origin of biological monomers
2. the origin of biological polymers
3. the evolution from molecules to cells

Bernal suggested that evolution commenced between stages 1 and 2. Bernal regarded the third stage – discovering methods by which biological reactions were incorporated behind a cell's boundary – as the most difficult. Modern work on the way that cell membranes self-assemble, and the work on micropores in various substrates may be a halfway house towards the development of independent free-living cells.

The chemical processes that took place on the early Earth are called *chemical evolution*. Since the end of the nineteenth century, 'evolutive abiogenesis' means increasing complexity and evolution of matter from inert to living state.[10] Both Manfred Eigen and Sol Spiegelman demonstrated that evolution, including replication, variation, and natural selection, can occur in populations of molecules as well as in organisms. Spiegelman took advantage of natural selection to synthesize the Spiegelman Monster, which had a genome with just 218 nucleotide bases, having deconstructively evolved from a 4500-base bacterial RNA. Eigen built on Spiegelman's work and produced a similar system further degraded to just 48 or 54 nucleotides – the minimum required for the binding of the replication enzyme.

Following on from chemical evolution came the initiation of biological evolution, which led to the first cells. No one has yet synthesized a "protocell" using simple components with the necessary properties of life (the so-called "bottom-up-approach"). Without such a proof-of-principle, explanations have tended to focus on chemosynthesis. However, some researchers work in this field, notably Steen Rasmussen and Jack W. Szostak. Others have argued that a "top-down approach" is more feasible. One such approach, successfully attempted by Craig Venter and others at J. Craig Venter Institute, involves engineering existing prokaryotic cells with progressively fewer genes, attempting to discern at which point the most minimal requirements for life are reached.[11]

The NASA strategy on abiogenesis states that it is necessary to identify interactions, intermediary structures and functions, energy sources, and environmental factors that contributed to the diversity, selection, and replication of evolvable macromolecular systems. Emphasis must continue to map the chemical landscape of potential primordial informational polymers. The advent of polymers that could replicate, store genetic information, and exhibit properties subject to selection likely was a critical step in the emergence of prebiotic chemical evolution.

Chemical origin of organic molecules

The elements, except for hydrogen and helium, ultimately derive from stellar nucleosynthesis. On 12 October 2016, astronomers reported that the very basic chemical ingredients of life — the carbon-hydrogen molecule (CH, or methylidyne radical), the carbon-hydrogen positive ion (CH+) and the carbon ion (C+) — are largely the result of ultraviolet light from stars, rather than other forms of radiation from supernovae and young stars, as thought earlier. Complex molecules, including organic molecules, form naturally both in space and on planets. There are two possible sources of organic molecules on the early Earth:

1. Terrestrial origins – organic molecule synthesis driven by impact shocks or by other energy sources (such as UV light, redox coupling, or electrical discharges; e.g., Miller's experiments)
2. Extraterrestrial origins – formation of organic molecules in interstellar dust clouds, which rain down on planets.[12] (See pseudo-panspermia)

Based on recent computer model studies, the complex organic molecules necessary for life may have formed in the protoplanetary disk of dust grains surrounding the Sun before the formation of the Earth. According to the computer studies, this same process may also occur around other stars that acquire planets. (Also see Extraterrestrial organic molecules).

Estimates of the production of organics from these sources suggest that the Late Heavy Bombardment before 3.5 Ga within the early atmosphere made available quantities of organics comparable to those produced by terrestrial sources.

It has been estimated that the Late Heavy Bombardment may also have effectively sterilized the Earth's surface to a depth of tens of metres. If life evolved deeper than this, it would have also been shielded from the early high levels of ultraviolet radiation from the T Tauri stage of the Sun's evolution. Simulations of geothermically heated oceanic crust yield far more organics than those found in the Miller-Urey experiments (see below). In the deep hydrothermal

Phylogenetic Tree of Life

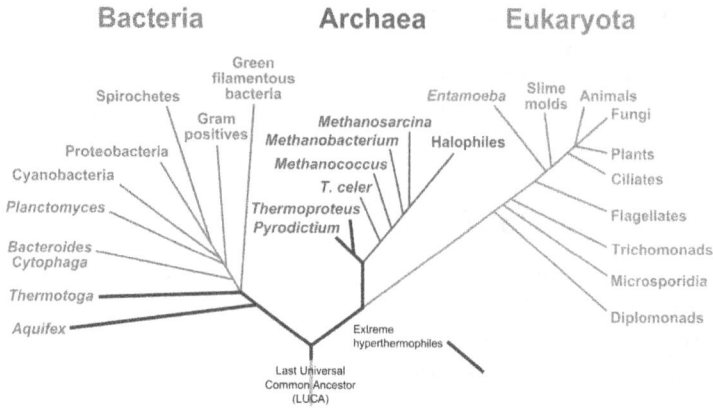

Figure 5: *A cladogram demonstrating extreme hyper-
thermophiles at the base of the phylogenetic tree of life.*

vents, Everett Shock has found "there is an enormous thermodynamic drive
to form organic compounds, as seawater and hydrothermal fluids, which are
far from equilibrium, mix and move towards a more stable state." Shock has
found that the available energy is maximized at around 100 – 150 degrees Cel-
sius, precisely the temperatures at which the hyperthermophilic bacteria and
thermoacidophilic archaea have been found, at the base of the phylogenetic
tree of life closest to the Last Universal Common Ancestor (LUCA).

The accumulation and concentration of organic molecules on a planetary surface is also considered an essential early step for the origin of life. Identifying and understanding the mechanisms that led to the production of prebiotic molecules in various environments is critical for establishing the inventory
of ingredients from which life originated on Earth, assuming that the abiotic
production of molecules ultimately influenced the selection of molecules from
which life emerged.

Chemical synthesis

While features of self-organization and self-replication are often considered the
hallmark of living systems, there are many instances of abiotic molecules ex-
hibiting such characteristics under proper conditions. Stan Palasek suggested
based on a theoretical model that self-assembly of ribonucleic acid (RNA)
molecules can occur spontaneously due to physical factors in hydrothermal

vents. Virus self-assembly within host cells has implications for the study of the origin of life, as it lends further credence to the hypothesis that life could have started as self-assembling organic molecules.

Multiple sources of energy were available for chemical reactions on the early Earth. For example, heat (such as from geothermal processes) is a standard energy source for chemistry. Other examples include sunlight and electrical discharges (lightning), among others. Computer simulations also suggest that cavitation in primordial water reservoirs such as breaking sea waves, streams and oceans can potentially lead to the synthesis of biogenic compounds.[13] Unfavourable reactions can also be driven by highly favourable ones, as in the case of iron-sulfur chemistry. For example, this was probably important for carbon fixation (the conversion of carbon from its inorganic form to an organic one).[14] Carbon fixation via iron-sulfur chemistry is highly favourable, and occurs at neutral pH and 100 °C (212 °F). Iron-sulfur surfaces, which are abundant near hydrothermal vents, are also capable of producing small amounts of amino acids and other biological metabolites.

As early as the 1860s, experiments have demonstrated that biologically relevant molecules can be produced from interaction of simple carbon sources with abundant inorganic catalysts. In particular, experiments by Butlerov (the formose reaction) showed that tetroses, pentoses, and hexoses are produced when formaldehyde is heated under basic conditions with divalent metal ions like calcium. The reaction was scrutinized and subsequently proposed to be autocatalytic by Breslow in 1959. Similar experiments (see below) demonstrate that nucleobases like guanine and adenine could be synthesized from simple carbon and nitrogen sources like hydrogen cyanide and ammonia.

Formamide produces all four ribonucleotides and other biological molecules when warmed in the presence of various terrestrial minerals. Formamide is ubiquitous in the Universe, produced by the reaction of water and hydrogen cyanide (HCN). It has several advantages as a biotic precursor, including the ability to easily become concentrated through the evaporation of water. Although HCN is poisonous, it only affects aerobic organisms (eukaryotes and aerobic bacteria), which did not yet exist. It can play roles in other chemical processes as well, such as the synthesis of the amino acid glycine.

In 1961, it was shown that the nucleic acid purine base adenine can be formed by heating aqueous ammonium cyanide solutions. Other pathways for synthesizing bases from inorganic materials were also reported. Leslie E. Orgel and colleagues have shown that freezing temperatures are advantageous for the synthesis of purines, due to the concentrating effect for key precursors such as hydrogen cyanide. Research by Stanley L. Miller and colleagues suggested that while adenine and guanine require freezing conditions for synthesis, cytosine and uracil may require boiling temperatures. Research by the Miller group notes the formation of seven different amino acids and 11 types of nucleobases in ice when ammonia and cyanide were left in a freezer from 1972 to 1997. Other work demonstrated the formation of s-triazines (alternative nucleobases), pyrimidines (including cytosine and uracil), and adenine from urea solutions subjected to freeze-thaw cycles under a reductive atmosphere (with spark discharges as an energy source). The explanation given for the unusual speed of these reactions at such a low temperature is eutectic freezing. As an ice crystal forms, it stays pure: only molecules of water join the growing crystal, while impurities like salt or cyanide are excluded. These impurities become crowded in microscopic pockets of liquid within the ice, and this crowding causes the molecules to collide more often. Mechanistic exploration using quantum chemical methods provide a more detailed understanding of some of the chemical processes involved in chemical evolution, and a partial answer to the fundamental question of molecular biogenesis.

At the time of the Miller–Urey experiment, scientific consensus was that the early Earth had a reducing atmosphere with compounds relatively rich in hydrogen and poor in oxygen (e.g., CH_4 and NH_3 as opposed to CO_2 and nitrogen dioxide (NO_2)). However, current scientific consensus describes the primitive atmosphere as either weakly reducing or neutral (see also Oxygen Catastrophe). Such an atmosphere would diminish both the amount and variety of amino acids that could be produced, although studies that include iron and carbonate minerals (thought present in early oceans) in the experimental conditions have again produced a diverse array of amino acids. Other scientific research has focused on two other potential reducing environments: outer space and deep-sea thermal vents.

The spontaneous formation of complex polymers from abiotically generated monomers under the conditions posited by the "soup" theory is not at all a straightforward process. Besides the necessary basic organic monomers, compounds that would have prohibited the formation of polymers were also formed in high concentration during the Miller–Urey and Joan Oró experiments. The Miller–Urey experiment, for example, produces many substances that would react with the amino acids or terminate their coupling into peptide chains.

A research project completed in March 2015 by John D. Sutherland and others found that a network of reactions beginning with hydrogen cyanide and hydrogen sulfide, in streams of water irradiated by UV light, could produce the chemical components of proteins and lipids, as well as those of RNA, while not producing a wide range of other compounds. The researchers used the term "cyanosulfidic" to describe this network of reactions.

Autocatalysis

Autocatalysts are substances that catalyze the production of themselves and therefore are "molecular replicators." The simplest self-replicating chemical systems are autocatalytic, and typically contain three components: a product molecule and two precursor molecules. The product molecule joins together the precursor molecules, which in turn produce more product molecules from more precursor molecules. The product molecule catalyzes the reaction by providing a complementary template that binds to the precursors, thus bringing them together. Such systems have been demonstrated both in biological macromolecules and in small organic molecules. Systems that do not proceed by template mechanisms, such as the self-reproduction of micelles and vesicles, have also been observed.

It has been proposed that life initially arose as autocatalytic chemical networks. British ethologist Richard Dawkins wrote about autocatalysis as a potential explanation for the origin of life in his 2004 book *The Ancestor's Tale*. In his book, Dawkins cites experiments performed by Julius Rebek Jr. and his colleagues in which they combined amino adenosine and pentafluorophenyl esters with the autocatalyst amino adenosine triacid ester (AATE). One product was a variant of AATE, which catalyzed the synthesis of themselves. This experiment demonstrated the possibility that autocatalysts could exhibit competition within a population of entities with heredity, which could be interpreted as a rudimentary form of natural selection.

In the early 1970s, Manfred Eigen and Peter Schuster examined the transient stages between the molecular chaos and a self-replicating hypercycle in a prebiotic soup. In a hypercycle, the information storing system (possibly RNA) produces an enzyme, which catalyzes the formation of another information system, in sequence until the product of the last aids in the formation of the first information system. Mathematically treated, hypercycles could create quasispecies, which through natural selection entered into a form of Darwinian evolution. A boost to hypercycle theory was the discovery of ribozymes capable of catalyzing their own chemical reactions. The hypercycle theory requires the existence of complex biochemicals, such as nucleotides, which do not form under the conditions proposed by the Miller–Urey experiment.

Geoffrey W. Hoffmann has shown that an early error-prone translation machinery can be stable against an error catastrophe of the type that had been envisaged as problematical for the origin of life, and was known as "Orgel's paradox".

Hoffmann has furthermore argued that a complex nucleation event as the origin of life involving both polypeptides and nucleic acid is compatible with the time and space available in the primitive oceans of Earth Hoffmann suggests that volcanic ash may provide the many random shapes needed in the postulated complex nucleation event. This aspect of the theory can be tested experimentally.

Homochirality

Homochirality refers to a geometric uniformity of some materials composed of chiral units. Chiral refers to nonsuperimposable 3D forms that are mirror images of one another, as are left and right hands. Living organisms use molecules that have the same chirality ("handedness"): with almost no exceptions, amino acids are left-handed while nucleotides and sugars are right-handed. Chiral molecules can be synthesized, but in the absence of a chiral source or a chiral catalyst, they are formed in a 50/50 mixture of both enantiomers (called a racemic mixture). Known mechanisms for the production of non-racemic mixtures from racemic starting materials include: asymmetric physical laws, such as the electroweak interaction; asymmetric environments, such as those caused by circularly polarized light, quartz crystals, or the Earth's rotation, statistical fluctuations during racemic synthesis,[15] and spontaneous symmetry breaking.

Once established, chirality would be selected for. A small bias (enantiomeric excess) in the population can be amplified into a large one by asymmetric autocatalysis, such as in the Soai reaction. In asymmetric autocatalysis, the catalyst is a chiral molecule, which means that a chiral molecule is catalyzing its own production. An initial enantiomeric excess, such as can be produced by polarized light, then allows the more abundant enantiomer to outcompete the other.

Clark has suggested that homochirality may have started in outer space, as the studies of the amino acids on the Murchison meteorite showed that L-alanine is more than twice as frequent as its D form, and L-glutamic acid was more than three times prevalent than its D counterpart. Various chiral crystal surfaces can also act as sites for possible concentration and assembly of chiral monomer units into macromolecules. Compounds found on meteorites suggest that the chirality of life derives from abiogenic synthesis, since amino acids from meteorites show a left-handed bias, whereas sugars show a predominantly right-handed bias, the same as found in living organisms.

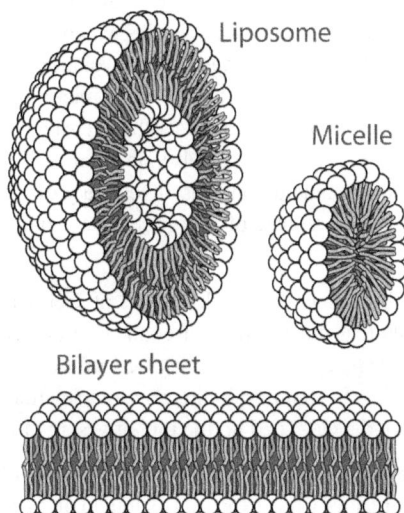

Figure 6: *The three main structures phospholipids form spontaneously in solution: the liposome (a closed bilayer), the micelle and the bilayer.*

Self-enclosement, reproduction, duplication and the RNA world

Protocells

A protocell is a self-organized, self-ordered, spherical collection of lipids proposed as a stepping-stone to the origin of life. A central question in evolution is how simple protocells first arose and differed in reproductive contribution to the following generation driving the evolution of life. Although a functional protocell has not yet been achieved in a laboratory setting, there are scientists who think the goal is well within reach.

Self-assembled vesicles are essential components of primitive cells. The second law of thermodynamics requires that the universe move in a direction in which entropy increases, yet life is distinguished by its great degree of organization. Therefore, a boundary is needed to separate life processes from non-living matter. Researchers Irene A. Chen and Jack W. Szostak amongst others, suggest that simple physicochemical properties of elementary protocells can give rise to essential cellular behaviours, including primitive forms of differential reproduction competition and energy storage. Such cooperative interactions between the membrane and its encapsulated contents could greatly simplify the transition from simple replicating molecules to true cells. Furthermore, competition for membrane molecules would favour stabilized membranes, suggesting a selective advantage for the evolution of cross-linked fatty acids and even the phospholipids of today. Such micro-encapsulation would allow for metabolism within the membrane, the exchange of small molecules but the prevention of passage of large substances across it. The main advantages of encapsulation include the increased solubility of the contained cargo within the capsule and the storage of energy in the form of a electrochemical gradient.

A 2012 study led by Armen Y. Mulkidjanian of Germany's University of Osnabrück, suggests that inland pools of condensed and cooled geothermal vapour have the ideal characteristics for the origin of life. Scientists confirmed in 2002 that by adding a montmorillonite clay to a solution of fatty acid micelles (lipid spheres), the clay sped up the rate of vesicles formation 100-fold.

Another protocell model is the Jeewanu. First synthesized in 1963 from simple minerals and basic organics while exposed to sunlight, it is still reported to have some metabolic capabilities, the presence of semipermeable membrane, amino acids, phospholipids, carbohydrates and RNA-like molecules. However, the nature and properties of the Jeewanu remains to be clarified.

Electrostatic interactions induced by short, positively charged, hydrophobic peptides containing 7 amino acids in length or fewer, can attach RNA to a vesicle membrane, the basic cell membrane.[16]

RNA world

The RNA world hypothesis describes an early Earth with self-replicating and catalytic RNA but no DNA or proteins. It is generally accepted that current life on Earth descends from an RNA world,[17] although RNA-based life may not have been the first life to exist. This conclusion is drawn from many independent lines of evidence, such as the observations that RNA is central to the translation process and that small RNAs can catalyze all of the chemical groups and information transfers required for life. The structure of the ribosome has been called the "smoking gun," as it showed that the ribosome is a ribozyme, with a central core of RNA and no amino acid side chains within 18 angstroms of the active site where peptide bond formation is catalyzed. The concept of the RNA world was first proposed in 1962 by Alexander Rich, and the term was coined by Walter Gilbert in 1986.

Possible precursors for the evolution of protein synthesis include a mechanism to synthesize short peptide cofactors or form a mechanism for the duplication of RNA. It is likely that the ancestral ribosome was composed entirely of RNA, although some roles have since been taken over by proteins. Major remaining

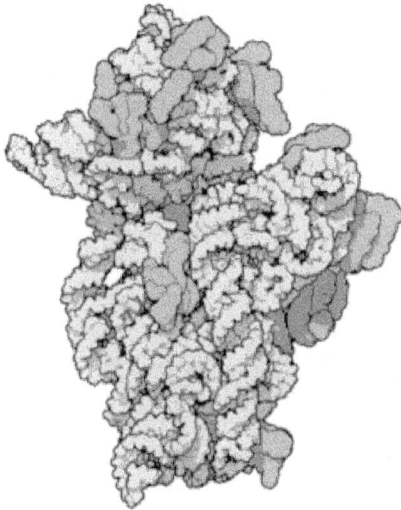

Figure 7: *Molecular structure of the ribosome 30S subunit from Thermus thermophilus. Proteins are shown in blue and the single RNA chain in orange.*

questions on this topic include identifying the selective force for the evolution of the ribosome and determining how the genetic code arose.

Eugene Koonin said, "Despite considerable experimental and theoretical effort, no compelling scenarios currently exist for the origin of replication and translation, the key processes that together comprise the core of biological systems and the apparent pre-requisite of biological evolution. The RNA World concept might offer the best chance for the resolution of this conundrum but so far cannot adequately account for the emergence of an efficient RNA replicase or the translation system. The MWO [Ed.: "many worlds in one"] version of the cosmological model of eternal inflation could suggest a way out of this conundrum because, in an infinite multiverse with a finite number of distinct macroscopic histories (each repeated an infinite number of times), emergence of even highly complex systems by chance is not just possible but inevitable."

Viral origins

Recent evidence for a "virus first" hypothesis, which may support theories of the RNA world, has been suggested. One of the difficulties for the study of the origins of viruses is their high rate of mutation; this is particularly the case in RNA retroviruses like HIV. A 2015 study compared protein fold structures across different branches of the tree of life, where researchers can reconstruct the evolutionary histories of the folds and of the organisms whose genomes code for those folds. They argue that protein folds are better markers of ancient events as their three-dimensional structures can be maintained even as the sequences that code for those begin to change. Thus, the viral protein repertoire retain traces of ancient evolutionary history that can be recovered using advanced bioinformatics approaches. Those researchers think that "the prolonged pressure of genome and particle size reduction eventually reduced virocells into modern viruses (identified by the complete loss of cellular makeup), meanwhile other coexisting cellular lineages diversified into modern cells. The data suggest that viruses originated from ancient cells that co-existed with the ancestors of modern cells. These ancient cells likely contained segmented RNA genomes.

RNA synthesis and replication

A number of hypotheses of formation of RNA have been put forward. As of 1994[4], there were difficulties in the explanation of the abiotic synthesis of the nucleotides cytosine and uracil. Subsequent research has shown possible routes of synthesis; for example, formamide produces all four ribonucleotides and other biological molecules when warmed in the presence of various terrestrial minerals. Early cell membranes could have formed spontaneously from proteinoids, which are protein-like molecules produced when amino acid solutions are heated while in the correct concentration of aqueous solution. These are seen to form micro-spheres which are observed to behave similarly to membrane-enclosed compartments. Other possible means of producing more complicated organic molecules include chemical reactions that take place on clay substrates or on the surface of the mineral pyrite.

Factors supporting an important role for RNA in early life include its ability to act both to store information and to catalyze chemical reactions (as a ribozyme); its many important roles as an intermediate in the expression of and maintenance of the genetic information (in the form of DNA) in modern organisms; and the ease of chemical synthesis of at least the components of the RNA molecule under the conditions that approximated the early Earth. Relatively short RNA molecules have been synthesized, capable of replication. Such replicase RNA, which functions as both code and catalyst provides its own template upon which copying can occur. Jack W. Szostak has shown that certain catalytic RNAs can join smaller RNA sequences together, creating the potential for self-replication. If these conditions were present, Darwinian natural selection would favour the proliferation of such autocatalytic sets, to which further functionalities could be added. Such autocatalytic systems of RNA capable of self-sustained replication have been identified. The RNA replication systems, which include two ribozymes that catalyze each other's synthesis, showed a doubling time of the product of about one hour, and were subject to natural selection under the conditions that existed in the experiment. In evolutionary competition experiments, this led to the emergence of new systems which replicated more efficiently. This was the first demonstration of evolutionary adaptation occurring in a molecular genetic system.

Depending on the definition, life started when RNA chains began to self-replicate, initiating the three mechanisms of Darwinian selection: heritability, variation of type, and differential reproductive output. The fitness of an RNA replicator (its per capita rate of increase) would likely be a function of its intrinsic adaptive capacities, determined by its nucleotide sequence, and the availability of resources. The three primary adaptive capacities may have been: (1) replication with moderate fidelity, giving rise to both heritability while allowing variation of type, (2) resistance to decay, and (3) acquisition of process resources. These capacities would have functioned by means of the folded configurations of the RNA replicators resulting from their nucleotide sequences.

Carl Zimmer has speculated that the chemical conditions, including the presence of boron, molybdenum and oxygen needed for the initial production of RNA, may have been better on early Mars than on early Earth. If so, life-suitable molecules originating on Mars may have later migrated to Earth via meteor ejections.

Pre-RNA world

It is possible that a different type of nucleic acid, such as PNA, TNA or GNA, was the first to emerge as a self-reproducing molecule, only later replaced by RNA. Larralde et al., say that "the generally accepted prebiotic synthesis of ribose, the formose reaction, yields numerous sugars without any selectivity." and they conclude that their "results suggest that the backbone of the first genetic material could not have contained ribose or other sugars because of their instability." The ester linkage of ribose and phosphoric acid in RNA is known to be prone to hydrolysis.

Pyrimidine ribonucleosides and their respective nucleotides have been prebiotically synthesized by a sequence of reactions which by-pass the free sugars, and are assembled in a stepwise fashion by using nitrogenous or oxygenous chemistries. Sutherland has demonstrated high yielding routes to cytidine and uridine ribonucleotides built from small 2 and 3 carbon fragments such as glycolaldehyde, glyceraldehyde or glyceraldehyde-3-phosphate, cyanamide and cyanoacetylene. One of the steps in this sequence allows the isolation of enantiopure ribose aminooxazoline if the enantiomeric excess of glyceraldehyde is

60% or greater. This can be viewed as a prebiotic purification step, where the said compound spontaneously crystallized out from a mixture of the other pentose aminooxazolines. Ribose aminooxazoline can then react with cyanoacetylene in a mild and highly efficient manner to give the alpha cytidine ribonucleotide. Photoanomerization with UV light allows for inversion about the 1' anomeric centre to give the correct beta stereochemistry. In 2009 they showed that the same simple building blocks allow access, via phosphate controlled nucleobase elaboration, to 2',3'-cyclic pyrimidine nucleotides directly, which are known to be able to polymerize into RNA. This paper also highlights the possibility for the photo-sanitization of the pyrimidine-2',3'-cyclic phosphates.

Origin of biological metabolism

Metabolism-like reactions could have occurred naturally in early oceans, before the first organisms evolved. Metabolism may predate the origin of life, which may have evolved from the chemical conditions in the earliest oceans. Reconstructions in laboratories show that some of these reactions can produce RNA, and some others resemble two essential reaction cascades of metabolism: glycolysis and the pentose phosphate pathway, that provide essential precursors for nucleic acids, amino acids and lipids. A study at the University of Düsseldorf created phylogenic trees based upon 6 million genes from bacteria and archaea, and identified 355 protein families that were probably present in the LUCA. They were based upon an anaerobic metabolism fixing carbon dioxide and nitrogen. It suggests that the LUCA evolved in an environment rich in hydrogen, carbon dioxide and iron.[18] Following are some observed discoveries and related hypotheses.

Iron–sulfur world

In the 1980s, Günter Wächtershäuser, encouraged and supported by Karl R. Popper, postulated his iron–sulfur world, a theory of the evolution of prebiotic chemical pathways as the starting point in the evolution of life. It systematically traces today's biochemistry to primordial reactions which provide alternative pathways to the synthesis of organic building blocks from simple gaseous compounds.

In contrast to the classical Miller experiments, which depend on external sources of energy (simulated lightning, ultraviolet irradiation), "Wächtershäuser systems" come with a built-in source of energy: sulfides of iron (iron pyrite) and other minerals. The energy released from redox reactions of these metal sulfides is available for the synthesis of organic molecules, and such systems may have evolved into autocatalytic sets constituting self-replicating, metabolically active entities predating the life forms known today. Experiments with such sulfides in an aqueous environment at 100 °C produced a relatively small yield of dipeptides (0.4% to 12.4%) and a smaller yield of tripeptides (0.10%) although under the same conditions, dipeptides were quickly broken down.

Several models reject the self-replication of a "naked-gene", postulating instead the emergence of a primitive metabolism providing a safe environment for the later emergence of RNA replication. The centrality of the Krebs cycle (citric acid cycle) to energy production in aerobic organisms, and in drawing in carbon dioxide and hydrogen ions in biosynthesis of complex organic chemicals, suggests that it was one of the first parts of the metabolism to evolve. Concordantly, geochemist Michael Russell has proposed that "the purpose of life is to hydrogenate carbon dioxide" (as part of a "metabolism-first," rather than a "genetics-first," scenario). Physicist Jeremy England of MIT has proposed that life was inevitable from general thermodynamic considerations: "... when a group of atoms is driven by an external source of energy (like the sun or chemical fuel) and surrounded by a heat bath (like the ocean or atmosphere), it will often gradually restructure itself in order to dissipate increasingly more energy. This could mean that under certain conditions, matter inexorably acquires the key physical attribute associated with life."

One of the earliest incarnations of this idea was put forward in 1924 with Oparin's notion of primitive self-replicating vesicles which predated the discovery of the structure of DNA. Variants in the 1980s and 1990s include Wächtershäuser's iron–sulfur world theory and models introduced by Christian de Duve based on the chemistry of thioesters. More abstract and theoretical arguments for the plausibility of the emergence of metabolism without the presence of genes include a mathematical model introduced by Freeman Dyson in the early 1980s and Stuart Kauffman's notion of collectively autocatalytic sets, discussed later that decade.

Orgel summarized his analysis by stating, "There is at present no reason to expect that multistep cycles such as the reductive citric acid cycle will self-organize on the surface of FeS/FeS_2 or some other mineral." It is possible that another type of metabolic pathway was used at the beginning of life. For example, instead of the reductive citric acid cycle, the "open" acetyl-CoA pathway (another one of the five recognized ways of carbon dioxide fixation in nature today) would be compatible with the idea of self-organization on a metal sulfide surface. The key enzyme of this pathway, carbon monoxide dehydrogenase/acetyl-CoA synthase, harbours mixed nickel-iron-sulfur clusters in its reaction centres and catalyzes the formation of acetyl-CoA (similar to acetylthiol) in a single step. There are increasing concerns, however, that prebiotic

thiolated and thioester compounds are thermodynamically and kinetically unfavourable to accumulate in presumed prebiotic conditions (i.e. hydrothermal
vents). It has also been proposed that cysteine and homocysteine may have reacted with nitriles resulting from the Stecker reaction, readily forming catalytic
thiol-reach poplypeptides.

Zn-world hypothesis

The Zn-world (zinc world) theory of Armen Y. Mulkidjanian is an extension
of Wächtershäuser's pyrite hypothesis. Wächtershäuser based his theory of the
initial chemical processes leading to informational molecules (RNA, peptides)
on a regular mesh of electric charges at the surface of pyrite that may have
facilitated the primeval polymerization by attracting reactants and arranging
them appropriately relative to each other. The Zn-world theory specifies and
differentiates further. Hydrothermal fluids rich in H_2S interacting with cold
primordial ocean (or Darwin's "warm little pond") water leads to the precipitation of metal sulfide particles. Oceanic vent systems and other hydrothermal
systems have a zonal structure reflected in ancient volcanogenic massive sulfide deposits (VMS) of hydrothermal origin. They reach many kilometres in
diameter and date back to the Archean Eon. Most abundant are pyrite (FeS_2),
chalcopyrite ($CuFeS_2$), and sphalerite (ZnS), with additions of galena (PbS)
and alabandite (MnS). ZnS and MnS have a unique ability to store radiation
energy, e.g. from UV light. During the relevant time window of the origins of
replicating molecules, the primordial atmospheric pressure was high enough
(>100 bar, about 100 atmospheres) to precipitate near the Earth's surface, and
UV irradiation was 10 to 100 times more intense than now; hence the unique
photosynthetic properties mediated by ZnS provided just the right energy conditions to energize the synthesis of informational and metabolic molecules and
the selection of photostable nucleobases.

The Zn-world theory has been further filled out with experimental and theoretical evidence for the ionic constitution of the interior of the first proto-cells
before archaea, bacteria and proto-eukaryotes evolved. Archibald Macallum
noted the resemblance of body fluids such as blood and lymph to seawater;
however, the inorganic composition of all cells differ from that of modern seawater, which led Mulkidjanian and colleagues to reconstruct the "hatcheries"
of the first cells combining geochemical analysis with phylogenomic scrutiny
of the inorganic ion requirements of universal components of modern cells.
The authors conclude that ubiquitous, and by inference primordial, proteins
and functional systems show affinity to and functional requirement for K^+,
Zn^{2+}, Mn^{2+}, and phosphate. Geochemical reconstruction shows that the ionic
composition conducive to the origin of cells could not have existed in what

Figure 8: *Deep-sea hydrothermal vent or black smoker*

we today call marine settings but is compatible with emissions of vapour-dominated zones of what we today call inland geothermal systems. Under the oxygen depleted, CO_2-dominated primordial atmosphere, the chemistry of water condensates and exhalations near geothermal fields would resemble the internal milieu of modern cells. Therefore, the precellular stages of evolution may have taken place in shallow "Darwin ponds" lined with porous silicate minerals mixed with metal sulfides and enriched in K^+, Zn^{2+}, and phosphorus compounds.[19]

Deep sea vent hypothesis

The deep sea vent, or alkaline hydrothermal vent, theory posits that life may have begun at submarine hydrothermal vents, William Martin and Michael Russell have suggested "that life evolved in structured iron monosulphide precipitates in a seepage site hydrothermal mound at a redox, pH, and temperature gradient between sulphide-rich hydrothermal fluid and iron(II)-containing waters of the Hadean ocean floor. The naturally arising, three-dimensional compartmentation observed within fossilized seepage-site metal sulphide precipitates indicates that these inorganic compartments were the precursors of cell walls and membranes found in free-living prokaryotes. The known capability of FeS and NiS to catalyze the synthesis of the acetyl-methylsulphide from carbon monoxide and methylsulphide, constituents of hydrothermal fluid, indicates that pre-biotic syntheses occurred at the inner surfaces of these metal-sulphide-walled compartments,..." These form where hydrogen-rich fluids emerge from below the sea floor, as a result of serpentinization of ultramafic olivine with seawater and a pH interface with carbon dioxide-rich ocean water. The vents form a sustained chemical energy source derived from redox reactions, in which electron donors (molecular hydrogen) react with electron acceptors (carbon dioxide); see Iron–sulfur world theory. These are highly exothermic reactions.[20]

Michael Russell demonstrated that alkaline vents created an abiogenic proton motive force (PMF) chemiosmotic gradient, in which conditions are ideal for an abiogenic hatchery for life. Their microscopic compartments "provide a natural means of concentrating organic molecules," composed of iron-sulfur minerals such as mackinawite, endowed these mineral cells with the catalytic properties envisaged by Wächtershäuser. This movement of ions across the membrane depends on a combination of two factors:

1. Diffusion force caused by concentration gradient—all particles including ions tend to diffuse from higher concentration to lower.
2. Electrostatic force caused by electrical potential gradient—cations like protons H+ tend to diffuse down the electrical potential, anions in the opposite direction.

These two gradients taken together can be expressed as an electrochemical gradient, providing energy for abiogenic synthesis. The proton motive force can be described as the measure of the potential energy stored as a combination of proton and voltage gradients across a membrane (differences in proton concentration and electrical potential).

Jack W. Szostak suggested that geothermal activity provides greater opportunities for the origination of life in open lakes where there is a buildup of minerals. In 2010, based on spectral analysis of sea and hot mineral water, Ignat Ignatov and Oleg Mosin demonstrated that life may have predominantly originated in hot mineral water. The hot mineral water that contains bicarbonate and calcium ions has the most optimal range. This case is similar to the origin of life in hydrothermal vents, but with bicarbonate and calcium ions in hot water. This water has a pH of 9–11 and is possible to have the reactions in seawater. According to Melvin Calvin, certain reactions of condensation-dehydration of amino acids and nucleotides in individual blocks of peptides and nucleic acids can take place in the primary hydrosphere with pH 9-11 at a later evolutionary stage. Some of these compounds like hydrocyanic acid (HCN) have been proven in the experiments of Miller. This is the environment in which the stromatolites have been created. David Ward of Montana State University described the formation of stromatolites in hot mineral water at the Yellowstone National Park. Stromatolites survive in hot mineral water and in proximity to areas with volcanic activity. Processes have evolved in the sea near geysers of hot mineral water. In 2011, Tadashi Sugawara from the University of Tokyo created a protocell in hot water.

Experimental research and computer modelling suggest that the surfaces of mineral particles inside hydrothermal vents have catalytic properties similar to those of enzymes and are able to create simple organic molecules, such as methanol (CH_3OH) and formic, acetic and pyruvic acid out of the dissolved CO_2 in the water.

The research reported above by William F. Martin in July 2016 supports the thesis that life arose at hydrothermal vents, that spontaneous chemistry in the Earth's crust driven by rock–water interactions at disequilibrium thermodynamically underpinned life's origin[21] and that the founding lineages of the archaea and bacteria were H2-dependent autotrophs that used CO2 as their terminal acceptor in energy metabolism. Martin suggests, based upon this evidence that LUCA "may have depended heavily on the geothermal energy of the vent to survive".[22]

Thermosynthesis

Today's bioenergetic process of fermentation is carried out by either the aforementioned citric acid cycle or the Acetyl-CoA pathway, both of which have been connected to the primordial Iron–sulfur world. In a different approach, the thermosynthesis hypothesis considers the bioenergetic process of chemiosmosis, which plays an essential role in cellular respiration and photosynthesis, more basal than fermentation: the ATP synthase enzyme, which sustains chemiosmosis, is proposed as the currently extant enzyme most closely related to the first metabolic process.

First, life needed an energy source to bring about the condensation reaction that yielded the peptide bonds of proteins and the phosphodiester bonds of RNA. In a generalization and thermal variation of the binding change mechanism of today's ATP synthase, the "first protein" would have bound substrates (peptides, phosphate, nucleosides, RNA 'monomers') and condensed them to a reaction product that remained bound until after a temperature change it was released by thermal unfolding.

The energy source under the thermosynthesis hypothesis was thermal cycling, the result of suspension of protocells in a convection current, as is plausible in a volcanic hot spring; the convection accounts for the self-organization and dissipative structure required in any origin of life model. The still ubiquitous role of thermal cycling in germination and cell division is considered a relic of primordial thermosynthesis.

By phosphorylating cell membrane lipids, this "first protein" gave a selective advantage to the lipid protocell that contained the protein. This protein also synthesized a library of many proteins, of which only a minute fraction had thermosynthesis capabilities. As proposed by Dyson, it propagated functionally: it made daughters with similar capabilities, but it did not copy itself. Functioning daughters consisted of different amino acid sequences.

Whereas the Iron–sulfur world identifies a circular pathway as the most simple, the thermosynthesis hypothesis does not even invoke a pathway: ATP synthase's binding change mechanism resembles a physical adsorption process that yields free energy, rather than a regular enzyme's mechanism, which decreases the free energy. It has been claimed that the emergence of cyclic systems of protein catalysts is implausible.

Other models

Nature timeline

Axis scale: billion years

👆

Also see: *Human timeline* and *Life timeline*

Clay hypothesis

Montmorillonite, an abundant clay, is a catalyst for the polymerization of RNA and for the formation of membranes from lipids. A model for the origin of life using clay was forwarded by Alexander Graham Cairns-Smith in 1985 and explored as a plausible mechanism by several scientists. The clay hypothesis postulates that complex organic molecules arose gradually on pre-existing, non-organic replication surfaces of silicate crystals in solution.

At the Rensselaer Polytechnic Institute, James P. Ferris' studies have also confirmed that clay minerals of montmorillonite catalyze the formation of RNA in aqueous solution, by joining nucleotides to form longer chains.

In 2007, Bart Kahr from the University of Washington and colleagues reported their experiments that tested the idea that crystals can act as a source of transferable information, using crystals of potassium hydrogen phthalate. "Mother" crystals with imperfections were cleaved and used as seeds to grow "daughter" crystals from solution. They then examined the distribution of imperfections in the new crystals and found that the imperfections in the mother crystals were reproduced in the daughters, but the daughter crystals also had many additional imperfections. For gene-like behaviour to be observed, the quantity of inheritance of these imperfections should have exceeded that of the mutations in the successive generations, but it did not. Thus Kahr concluded that the crystals "were not faithful enough to store and transfer information from one generation to the next."

Gold's "deep-hot biosphere" model

In the 1970s, Thomas Gold proposed the theory that life first developed not on the surface of the Earth, but several kilometres below the surface. It is claimed that discovery of microbial life below the surface of another body in our Solar System would lend significant credence to this theory. Thomas Gold also asserted that a trickle of food from a deep, unreachable, source is needed for survival because life arising in a puddle of organic material is likely to consume all of its food and become extinct. Gold's theory is that the flow of such food is due to out-gassing of primordial methane from the Earth's mantle; more conventional explanations of the food supply of deep microbes (away from sedimentary carbon compounds) is that the organisms subsist on hydrogen released by an interaction between water and (reduced) iron compounds in rocks.

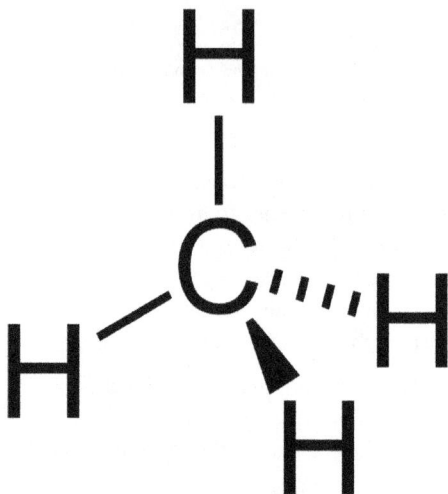

Figure 9: *Methane is one of the simplest organic compounds*

Panspermia

Panspermia is the hypothesis that life exists throughout the universe, distributed by meteoroids, asteroids, comets, planetoids,[23] and, also, by spacecraft in the form of unintended contamination by microorganisms.[24]

The panspermia hypothesis does not attempt to explain how life first originated, but merely shifts it to another planet or a comet. The advantage of an extraterrestrial origin of primitive life is that life is not required to have formed on each planet it occurs on, but rather in a single location, and then spread about the galaxy to other star systems via cometary and/or meteorite impact. Evidence to support the hypothesis is scant, but it finds support in studies of Martian meteorites found in Antarctica and in studies of extremophile microbes' survival in outer space tests. (See also: List of microorganisms tested in outer space.)

Extraterrestrial organic molecules

An organic compound is any member of a large class of gaseous, liquid, or solid chemicals whose molecules contain carbon. Carbon is the fourth most abundant element in the Universe by mass after hydrogen, helium, and oxygen. Carbon is abundant in the Sun, stars, comets, and in the atmospheres of most planets. Organic compounds are relatively common in space, formed by "factories of complex molecular synthesis" which occur in molecular clouds and circumstellar envelopes, and chemically evolve after reactions are initiated mostly by ionizing radiation. Based on computer model studies, the complex organic molecules necessary for life may have formed on dust grains in the protoplanetary disk surrounding the Sun before the formation of the Earth. According to the computer studies, this same process may also occur around other stars that acquire planets.

Observations suggest that the majority of organic compounds introduced on Earth by interstellar dust particles are considered principal agents in the formation of complex molecules, thanks to their peculiar surface-catalytic activities.[25] Studies reported in 2008, based on $^{12}C/^{13}C$ isotopic ratios of organic compounds found in the Murchison meteorite, suggested that the RNA component uracil and related molecules, including xanthine, were formed extraterrestrially. On 8 August 2011, a report based on NASA studies of meteorites found on Earth was published suggesting DNA components (adenine, guanine and related organic molecules) were made in outer space. Scientists also found that the cosmic dust permeating the universe contains complex organics ("amorphous organic solids with a mixed aromatic–aliphatic structure") that could be created naturally, and rapidly, by stars.[26] Sun Kwok of The University of Hong Kong suggested that these compounds may have been related to the development of life on Earth said that "If this is the case, life on Earth may have had an easier time getting started as these organics can serve as basic ingredients for life."

Glycolaldehyde, the first example of an interstellar sugar molecule, was detected in the star-forming region near the centre of our galaxy. It was discovered in 2000 by Jes Jørgensen and Jan M. Hollis. In 2012, Jørgensen's team reported the detection of glycolaldehyde in a distant star system. The molecule was found around the protostellar binary IRAS 16293-2422 400 light years from Earth. Glycolaldehyde is needed to form RNA, which is similar in function to DNA. These findings suggest that complex organic molecules may form in stellar systems prior to the formation of planets, eventually arriving on young planets early in their formation. Because sugars are associated with both metabolism and the genetic code, two of the most basic aspects of life, it is thought the discovery of extraterrestrial sugar increases the likelihood that life may exist elsewhere in our galaxy.

NASA announced in 2009 that scientists had identified another fundamental chemical building block of life in a comet for the first time, glycine, an amino acid, which was detected in material ejected from comet Wild 2 in 2004 and grabbed by NASA's *Stardust* probe. Glycine has been detected in meteorites before. Carl Pilcher, who leads the NASA Astrobiology Institute commented that "The discovery of glycine in a comet supports the idea that

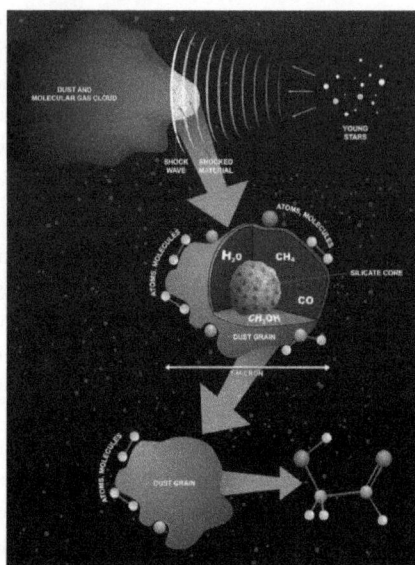

Figure 10: *Formation of glycolaldehyde in stardust*

the fundamental building blocks of life are prevalent in space, and strengthens the argument that life in the universe may be common rather than rare."
Comets are encrusted with outer layers of dark material, thought to be a tar-like substance composed of complex organic material formed from simple carbon compounds after reactions initiated mostly by ionizing radiation. It is possible that a rain of material from comets could have brought significant quantities of such complex organic molecules to Earth. Amino acids which were formed extraterrestrially may also have arrived on Earth via comets. It is estimated that during the Late Heavy Bombardment, meteorites may have delivered up to five million tons of organic prebiotic elements to Earth per year.

Polycyclic aromatic hydrocarbons (PAH) are the most common and abundant of the known polyatomic molecules in the observable universe, and are considered a likely constituent of the primordial sea. In 2010, PAHs, along with fullerenes (or "buckyballs"), have been detected in nebulae. In March 2015, NASA scientists reported that, for the first time, complex DNA and RNA organic compounds of life, including uracil, cytosine and thymine, have been formed in the laboratory under outer space conditions, using starting chemicals, such as pyrimidine, found in meteorites. Pyrimidine, like PAHs, the most carbon-rich chemical found in the Universe, may have been formed in red giant stars or in interstellar dust and gas clouds. A group of Czech scientists

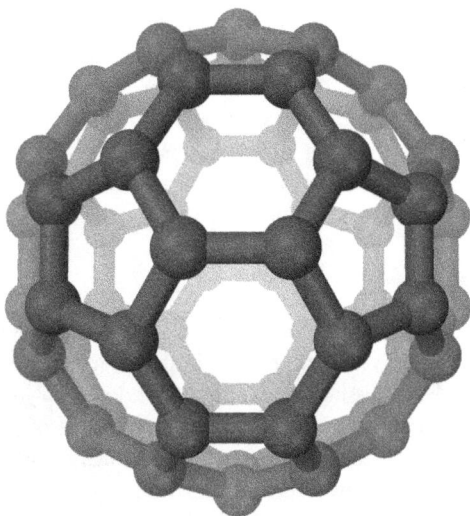

Figure 11: *The C_{60} buckyball is a complex molecule that has been detected in nebulae.*

reported that all four RNA-bases may be synthesized from formamide in the course of high-energy density events like extraterrestrial impacts.

Lipid world

The lipid world theory postulates that the first self-replicating object was lipid-like. It is known that phospholipids form lipid bilayers in water while under agitation—the same structure as in cell membranes. These molecules were not present on early Earth, but other amphiphilic long-chain molecules also form membranes. Furthermore, these bodies may expand (by insertion of additional lipids), and under excessive expansion may undergo spontaneous splitting which preserves the same size and composition of lipids in the two progenies. The main idea in this theory is that the molecular composition of the lipid bodies is the preliminary way for information storage, and evolution led to the appearance of polymer entities such as RNA or DNA that may store information favourably. Studies on vesicles from potentially pre-biotic amphiphiles have so far been limited to systems containing one or two types of amphiphiles. This in contrast to the output of simulated prebiotic chemical reactions, which typically produce very heterogeneous mixtures of compounds. Within the hypothesis of a lipid bilayer membrane composed of a mixture of various distinct amphiphilic compounds there is the opportunity of a huge number of theoretically possible combinations in the arrangements of these amphiphiles in the membrane. Among all these potential combinations, a specific local arrangement of the membrane would have favoured the constitution of a hypercycle, actually a positive feedback composed of two mutual catalysts represented by a membrane site and a specific compound trapped in the vesicle. Such site/compound pairs are transmissible to the daughter vesicles leading to the emergence of distinct lineages of vesicles which would have allowed Darwinian natural selection.[27]

Polyphosphates

A problem in most scenarios of abiogenesis is that the thermodynamic equilibrium of amino acid versus peptides is in the direction of separate amino acids. What has been missing is some force that drives polymerization. The resolution of this problem may well be in the properties of polyphosphates. Polyphosphates are formed by polymerization of ordinary monophosphate ions PO_4^{-3}. Several mechanisms of organic molecule synthesis have been investigated. Polyphosphates cause polymerization of amino acids into peptides. They are also logical precursors in the synthesis of such key biochemical compounds as adenosine triphosphate (ATP). A key issue seems to be that calcium reacts with soluble phosphate to form insoluble calcium phosphate (apatite), so some plausible mechanism must be found to keep calcium ions from causing precipitation of phosphate. There has been much work on this topic over the years, but an interesting new idea is that meteorites may have introduced reactive phosphorus species on the early Earth.

PAH world hypothesis

Polycyclic aromatic hydrocarbons (PAH) are known to be abundant in the universe,[28] including in the interstellar medium, in comets, and in meteorites, and are some of the most complex molecules so far found in space.

Other sources of complex molecules have been postulated, including extraterrestrial stellar or interstellar origin. For example, from spectral analyses, organic molecules are known to be present in comets and meteorites. In 2004, a team detected traces of PAHs in a nebula.[29] In 2010, another team also detected PAHs, along with fullerenes, in nebulae. The use of PAHs has also been proposed as a precursor to the RNA world in the PAH world hypothesis.Wikipedia:Citation needed The Spitzer Space Telescope has detected a star, HH 46-IR, which is forming by a process similar to that by which the Sun formed. In the disk of material surrounding the star, there is a very large range of molecules, including cyanide compounds, hydrocarbons, and carbon monoxide. In September 2012, NASA scientists reported that PAHs, subjected to interstellar medium conditions, are transformed, through hydrogenation, oxygenation and hydroxylation, to more complex organics—"a step along the path toward amino acids and nucleotides, the raw materials of proteins and DNA, respectively." Further, as a result of these transformations, the PAHs lose their spectroscopic signature which could be one of the reasons "for the lack of PAH detection in interstellar ice grains, particularly the outer regions of cold, dense clouds or the upper molecular layers of protoplanetary disks."

NASA maintains a database for tracking PAHs in the universe. More than 20% of the carbon in the universe may be associated with PAHs, possible starting materials for the formation of life. PAHs seem to have been formed shortly after the Big Bang, are widespread throughout the universe, and are associated with new stars and exoplanets.

Radioactive beach hypothesis

Zachary Adam claims that tidal processes that occurred during a time when the Moon was much closer may have concentrated grains of uranium and other radioactive elements at the high-water mark on primordial beaches, where they may have been responsible for generating life's building blocks. According to computer models, a deposit of such radioactive materials could show the same self-sustaining nuclear reaction as that found in the Oklo uranium ore seam in Gabon. Such radioactive beach sand might have provided sufficient energy to generate organic molecules, such as amino acids and sugars from acetonitrile in water. Radioactive monazite material also has released soluble phosphate into the regions between sand-grains, making it biologically "accessible." Thus amino acids, sugars, and soluble phosphates might have been produced simultaneously, according to Adam. Radioactive actinides, left behind in some concentration by the reaction, might have formed part of organometallic complexes. These complexes could have been important early catalysts to living processes.

John Parnell has suggested that such a process could provide part of the "crucible of life" in the early stages of any early wet rocky planet, so long as the planet is large enough to have generated a system of plate tectonics which brings radioactive minerals to the surface. As the early Earth is thought to have had many smaller plates, it might have provided a suitable environment for such processes.

Thermodynamic dissipation

The 19th-century Austrian physicist Ludwig Boltzmann first recognized that the struggle for existence of living organisms was neither over raw material nor energy, but instead had to do with entropy production derived from the conversion of the solar spectrum into heat by these systems.[30] Boltzmann thus realized that living systems, like all irreversible processes, were dependent on the dissipation of a generalized chemical potential for their existence. In his book "What is Life", the 20th-century Austrian physicist Erwin Schrödinger[31] emphasized the importance of Boltzmann's deep insight into the irreversible thermodynamic nature of living systems, suggesting that this was the physics and chemistry behind the origin and evolution of life. However, irreversible processes, and much less living systems, could not be conveniently analyzed under this perspective until Lars Onsager,[32] and later Ilya Prigogine,[33] developed an elegant mathematical formalism for treating the "self-organization" of material under a generalized chemical potential. This formalism became known as Classical Irreversible Thermodynamics and Prigogine was awarded the Nobel Prize in Chemistry in 1977 "for his contributions to non-equilibrium thermodynamics, particularly the theory of dissipative structures". The analysis of Prigogine showed that if a system were left to evolve under an imposed external potential, material could spontaneously organize (lower its entropy) forming what he called "dissipative structures" which would increase the dissipation of the externally imposed potential (augment the global entropy production). Non-equilibrium thermodynamics has since been successfully applied to the analysis of living systems, from the biochemical production of ATP to optimizing bacterial metabolic pathways[34] to complete ecosystems.[35]

In his "Thermodynamic Dissipation Theory of the Origin and Evolution of Life",[36] Karo Michaelian has taken the insight of Boltzmann and the work of Prigogine to its ultimate consequences regarding the origin of life. This theory postulates that the hallmark of the origin and evolution of life is the microscopic dissipative structuring of organic pigments and their proliferation over the entire Earth surface. Present day life augments the entropy production of Earth in its solar environment by dissipating ultraviolet and visible photons into heat through organic pigments in water. This heat then catalyzes a host of secondary dissipative processes such as the water cycle, ocean and wind currents, hurricanes, etc. Michaelian argues that if the thermodynamic function of life today is to produce entropy through photon dissipation in organic pigments, then this probably was its function at its very beginnings. It turns out that both RNA and DNA when in water solution are very strong absorbers and extremely rapid dissipaters of ultraviolet light within the 230–290 nm wavelength (UV-C) region, which is a part of the Sun's spectrum that could have penetrated the prebiotic atmosphere.[37] In fact, not only RNA and DNA, but many fundamental molecules of life (those common to all three domains of life) are also pigments that absorb in the UV-C, and many of these also have a chemical affinity to RNA and DNA.[38] Nucleic acids may thus have acted as acceptor molecules to the UV-C photon excited antenna pigment donor molecules by providing an ultrafast channel for dissipation. Michaelian has shown using the formalism of non-linear irreversible thermodynamics that there would have existed during the Archean a thermodynamic imperative to the abiogenic UV-C photochemical synthesis and proliferation of these pigments over the entire Earth surface if they acted as catalysts to augment the dissipation of the solar photons. By the end of the Archean, with life-induced ozone dissipating UV-C light in the Earth's upper atmosphere, it would have become ever more improbable for a completely new life to emerge that didn't rely on the complex metabolic pathways already existing since now the free energy in the photons arriving at Earth's surface would have been insufficient for direct breaking and remaking of covalent bonds. It has been suggested, however, that such changes in the surface flux of ultraviolet radiation due to geophysical events affecting the atmosphere could have been what promoted the development of complexity in life based on existing metabolic pathways, for example during the Cambrian explosion

Many salient characteristics of the fundamental molecules of life (those found in all three domains) all point directly to the involvement of UV-C light in the dissipative structuring of incipient life. Some of the most difficult problems concerning the origin of life, such as enzyme-less replication of RNA and DNA, homochirality of the fundamental molecules, and the origin of information encoding in RNA and DNA, also find an explanation within the same dissipative thermodynamic framework by considering the probable existence of a relation between primordial replication and UV-C photon dissipation. Michaelian suggests that it is erroneous to expect to describe the emergence, proliferation, or even evolution, of life without overwhelming reference to entropy production through the dissipation of a generalized chemical potential, in particular, the prevailing solar photon flux.

Multiple genesis

Different forms of life with variable origin processes may have appeared quasi-simultaneously in the early history of Earth. The other forms may be extinct (having left distinctive fossils through their different biochemistry—e.g., hypothetical types of biochemistry). It has been proposed that:

> The first organisms were self-replicating iron-rich clays which fixed carbon dioxide into oxalic and other dicarboxylic acids. This system of replicating clays and their metabolic phenotype then evolved into the sulfide rich region of the hotspring acquiring the ability to fix nitrogen. Finally phosphate was incorporated into the evolving system which allowed the synthesis of nucleotides and phospholipids. If biosynthesis recapitulates biopoiesis, then the synthesis of amino acids preceded the synthesis of the purine and pyrimidine bases. Furthermore the polymerization of the amino acid thioesters into polypeptides preceded the directed polymerization of amino acid esters by polynucleotides.

Fluctuating hydrothermal pools on volcanic islands or proto-continents

Armid Mulkidjanian and co-authors think that the marine environments did not provide the ionic balance and composition universally found in cells, as well as of ions required by essential proteins and ribozymes found in virtually all living organisms, especially with respect to K^+/Na^+ ratio, Mn^{2+}, Zn^{2+} and phosphate concentrations. The only known environments that mimic the needed conditions on Earth are found in terrestrial hydrothermal pools fed by steam vents. Additionally, mineral deposits in these environments under an anoxic atmosphere would have suitable pH (as opposed to current pools in an oxygenated atmosphere), contain precipitates of sulfide minerals that block harmful UV radiation, have wetting/drying cycles that concentrate substrate solutions to concentrations amenable to spontaneous formation of polymers of nucleic acids, polyesters and depsipeptides, both by chemical reactions in the hydrothermal environment, as well as by exposure to UV light during transport from vents to adjacent pools. Their hypothesized pre-biotic environments are similar to the deep-oceanic vent environments most commonly hypothesized, but add additional components that help explain peculiarities found in reconstructions of the Last Universal Common Ancestor (LUCA) of all living organisms.

Bruce Damer and David Deamer have come to the conclusion that cell membranes cannot be formed in salty seawater, and must therefore have originated in freshwater. Before the continents formed, the only dry land on Earth would be volcanic islands, where rainwater would form ponds where lipids could form the first stages towards cell membranes. These predecessors of true cells are assumed to have behaved more like a superorganism rather than individual structures, where the porous membranes would house molecules which would leak out and enter other protocells. Only when true cells had evolved would they gradually adapt to saltier environments and enter the ocean.

Colín-García et al. (2016) discuss the advantages and disadvantages of hydrothermal vents as primitive environments. They mention the exergonic reactions in such systems could have been a source of free energy that promoted chemical reactions, additional to their high mineralogical diversity which implies the induction of important chemical gradients, thus favoring the interaction between electron donors and acceptors. Colín-García et al. (2016) also summarize a set of experiments proposed to test the role of hydrothermal vents in prebiotic synthesis.

Information theory

A theory that speaks to the origin of life on Earth and other rocky planets posits life as an information system in which information content grows because of selection. Life must start with minimum possible information, or minimum possible departure from thermodynamic equilibrium, and it requires thermo-dynamically free energy accessible by means of its information content. The most benign circumstances, minimum entropy variations with abundant free energy, suggest the pore space in the first few kilometres of the surface. Free energy is derived from the condensed products of the chemical reactions taking place in the cooling nebula.

Bibliography

<templatestyles src="Template:Refbegin/styles.css" />

- Altermann, Wladyslaw (2009). "From Fossils to Astrobiology – A Roadmap to Fata Morgana?"[39] (PDF). In Seckbach, Joseph; Walsh, Maud. *From Fossils to Astrobiology: Records of Life on Earth and the Search for Extraterrestrial Biosignatures*. Cellular Origin, Life in Extreme Habitats and Astrobiology. **12**. Dordrecht, the Netherlands; London: Springer Science+Business Media. ISBN 978-1-4020-8836-0. LCCN 2008933212[40]. Retrieved 2015-06-05.Wikipedia:Link rot
- Bada, Jeffrey L.; Lazcano, Antonio (2009). "The Origin of Life". In Ruse, Michael; Travis, Joseph. *Evolution: The First Four Billion Years*. Foreword by Edward O. Wilson. Cambridge, MA: Belknap Press of Harvard University Press. ISBN 978-0-674-03175-3. LCCN 2008030270[41]. OCLC 225874308[42].
- Barton, Nicholas H.; Briggs, Derek E. G.; Eisen, Jonathan A.; et al. (2007). *Evolution*. Cold Spring Harbor, NY: Cold Spring Harbor Laboratory Press. ISBN 978-0-87969-684-9. LCCN 2007010767[43]. OCLC 86090399[44].
- Bastian, H. Charlton (1871). *The Modes of Origin of Lowest Organisms*[45]. London; New York: Macmillan and Company. LCCN 11004276[46]. OCLC 42959303[47]. Retrieved 2015-06-06.
- Bernal, J. D. (1951). *The Physical Basis of Life*. London: Routledge & Kegan Paul. LCCN 51005794[48].
- Bernal, J. D. (1960). "The Problem of Stages in Biopoesis". In Florkin, M. *Aspects of the Origin of Life*. International Series of Monographs on Pure and Applied Biology. Oxford, UK; New York: Pergamon Press. ISBN 978-1-4831-3587-8. LCCN 60013823[49].

- Bernal, J. D. (1967) [Reprinted work by A. I. Oparin originally published 1924; Moscow: The Moscow Worker]. *The Origin of Life.* The Weidenfeld and Nicolson Natural History. Translation of Oparin by Ann Synge. London: Weidenfeld & Nicolson. LCCN 67098482[50].
- Bock, Gregory R.; Goode, Jamie A., eds. (1996). *Evolution of Hydrothermal Ecosystems on Earth (and Mars?).* Ciba Foundation Symposium. **202**. Chichester, UK; New York: John Wiley & Sons. ISBN 0-471-96509-X. LCCN 96031351[51].
- Bondeson, Jan (1999). *The Feejee Mermaid and Other Essays in Natural and Unnatural History.* Ithaca, NY: Cornell University Press. ISBN 0-8014-3609-5. LCCN 98038295[52].
- Bryson, Bill (2004). *A Short History of Nearly Everything.* London: Black Swan. ISBN 978-0-552-99704-1. OCLC 55589795[53].
- Calvin, Melvin (1969). *Chemical Evolution: Molecular Evolution Towards the Origin of Living Systems on the Earth and Elsewhere.* Oxford, UK: Clarendon Press. ISBN 0-19-855342-0. LCCN 70415289[54]. OCLC 25220[55].
- Chaichian, Masud; Rojas, Hugo Perez; Tureanu, Anca (2014). "Physics and Life". *Basic Concepts in Physics: From the Cosmos to Quarks.* Undergraduate Lecture Notes in Physics. Berlin; Heidelberg: Springer Berlin Heidelberg. doi: 10.1007/978-3-642-19598-3_12[56]. ISBN 978-3-642-19597-6. ISSN 2192-4791[57]. LCCN 2013950482[58]. OCLC 900189038[59].
- Chang, Thomas Ming Swi (2007). *Artificial Cells: Biotechnology, Nanomedicine, Regenerative Medicine, Blood Substitutes, Bioencapsulation, and Cell/Stem Cell Therapy.* Regenerative Medicine, Artificial Cells and Nanomedicine. **1**. Hackensack, NJ: World Scientific. ISBN 978-981-270-576-1. LCCN 2007013738[60]. OCLC 173522612[61].
- Dalrymple, G. Brent (2001). "The age of the Earth in the twentieth century: a problem (mostly) solved". In Lewis, C. L. E.; Knell, S. J. *The Age of the Earth: from 4004 BC to AD 2002.* Geological Society Special Publication. **190**. London: Geological Society of London. Bibcode: 2001GSLSP.190..205D[62]. doi: 10.1144/gsl.sp.2001.190.01.14[63]. ISBN 1-86239-093-2. ISSN 0305-8719[64]. LCCN 2003464816[65]. OCLC 48570033[66].
- Darwin, Charles (1887). Darwin, Francis, ed. *The Life and Letters of Charles Darwin, Including an Autobiographical Chapter.* **3** (3rd ed.). London: John Murray. OCLC 834491774[67].
- Davies, Geoffrey F. (2007). "Chapter 2.3 Dynamics of the Hadean and Archaean Mantle". In van Kranendonk, Martin J.; Smithies, R. Hugh; Bennett, Vickie C. *Earth's Oldest Rocks.* Developments in Precambrian Geology. **15**. Amsterdam, the Netherlands; Boston: Elsevier. doi:

10.1016/S0166-2635(07)15023-4[68]. ISBN 978-0-444-52810-0. LCCN 2009525003[69].

- Davies, Paul (1999). *The Fifth Miracle: The Search for the Origin of Life*. London: Penguin Books. ISBN 0-14-028226-2.
- Dawkins, Richard (1996). *The Blind Watchmaker* (Reissue with a new introduction ed.). New York: W. W. Norton & Company. ISBN 0-393-31570-3. LCCN 96229669[70]. OCLC 35648431[71].
- Dawkins, Richard (2004). *The Ancestor's Tale: A Pilgrimage to the Dawn of Evolution*. Boston, MA: Houghton Mifflin. ISBN 0-618-00583-8. LCCN 2004059864[72]. OCLC 56617123[73].
- Dobell, Clifford (1960) [Originally published 1932; New York: Harcourt, Brace & Company]. *Antony van Leeuwenhoek and His 'Little Animals'*. New York: Dover Publications. LCCN 60002548[74].
- Dyson, Freeman (1999). *Origins of Life* (Revised ed.). Cambridge, UK; New York: Cambridge University Press. ISBN 0-521-62668-4. LCCN 99021079[75].
- Eigen, M.; Schuster, P. (1979). *The Hypercycle: A Principle of Natural Self-Organization*. Berlin; New York: Springer-Verlag. ISBN 0-387-09293-5. LCCN 79001315[76]. OCLC 4665354[77].
- Fesenkov, V. G. (1959). "Some Considerations about the Primaeval State of the Earth". In Oparin, A. I.; et al. *The Origin of Life on the Earth*[78]. I.U.B. Symposium Series. **1**. Edited for the International Union of Biochemistry by Frank Clark and R. L. M. Synge (English-French-German ed.). London; New York: Pergamon Press. ISBN 978-1-4832-2240-0. LCCN 59012060[79]. Retrieved 2015-06-03. International Symposium on the Origin of Life on the Earth (held at Moscow, 19–24 August 1957)
- Hazen, Robert M. (2005). *Genesis: The Scientific Quest for Life's Origin*. Washington, D.C.: Joseph Henry Press. ISBN 0-309-09432-1. LCCN 2005012839[80]. OCLC 60321860[81].
- Huxley, Thomas Henry (1968) [Originally published 1897]. "VIII Biogenesis and Abiogenesis [1870]"[82]. *Discourses, Biological and Geological*. Collected Essays. **VIII** (Reprint ed.). New York: Greenwood Press. LCCN 70029958[83]. Retrieved 2014-05-19.
- Kauffman, Stuart (1993). *The Origins of Order: Self-Organization and Selection in Evolution*. New York: Oxford University Press. ISBN 978-0-19-507951-7. LCCN 91011148[84]. OCLC 23253930[85].
- Kauffman, Stuart (1995). *At Home in the Universe: The Search for Laws of Self-Organization and Complexity*. New York: Oxford University Press. ISBN 0-19-509599-5. LCCN 94025268[86].
- Klyce, Brig (22 January 2001). Kingsley, Stuart A.; Bhathal, Ragbir, eds. *Panspermia Asks New Questions*[87]. The Search for Extraterrestrial Intelligence (SETI) in the Optical Spectrum III[88]. **4273**. Bellingham,

WA: SPIE. doi: 10.1117/12.435366[89]. ISBN 0-8194-3951-7. LCCN
2001279159[90]. Retrieved 2015-06-09. Proceedings of the SPIE held at
San Jose, CA, 22–24 January 2001

- Lane, Nick (2009). *Life Ascending: The 10 Great Inventions of Evolution*
  (1st American ed.). New York: W. W. Norton & Company. ISBN 978-0-
  393-06596-1. LCCN 2009005046[91]. OCLC 286488326[92].

- Lankenau, Dirk-Henner (2011). "Two RNA Worlds: Toward the Origin
  of Replication, Genes, Recombination and Repair". In Egel, Richard;
  Lankenau, Dirk-Henner; Mulkidjanian,, Armen Y. *Origins of Life: The
  Primal Self-Organization*. Heidelberg: Springer. doi: 10.1007/978-
  3-642-21625-1[93]. ISBN 978-3-642-21624-4. LCCN 2011935879[94].
  OCLC 733245537[95].

- Lennox, James G. (2001). *Aristotle's Philosophy of Biology: Studies
  in the Origins of Life Science*. Cambridge Studies in Philosophy and
  Biology. Cambridge, UK; New York: Cambridge University Press.
  ISBN 0-521-65976-0. LCCN 00026070[96].

- Michod, Richard E. (1999). "Darwinian Dynamics: Evolutionary Transitions in Fitness and Individuality". Princeton, NJ: Princeton University
  Press. ISBN 0-691-02699-8. LCCN 98004166[97]. OCLC 38948118[98].

- Oparin, A. I. (1953) [Originally published 1938; New York: The Macmillan Company]. *The Origin of Life*. Translation and new introduction by
  Sergius Morgulis (2nd ed.). Mineola, NY: Dover Publications. ISBN 0-
  486-49522-1. LCCN 53010161[99].

- Orgel, Leslie E. (1987). "Evolution of the Genetic Apparatus: A Re-
  view". *Evolution of Catalytic Function*. Cold Spring Harbor Symposia on
  Quantitative Biology. 52. Cold Spring Harbor, NY: Cold Spring Harbor
  Laboratory Press. doi: 10.1101/SQB.1987.052.01.004[100]. ISBN 0-
  87969-054-2. OCLC 19850881[101]. "Proceedings of a symposium held at
  Cold Spring Harbor Laboratory in 1987"

- Raven, Peter H.; Johnson, George B. (2002). *Biology* (6th ed.). Boston,
  MA: McGraw-Hill. ISBN 0-07-112261-3. LCCN 2001030052[102].
  OCLC 45806501[103].

- Ross, Alexander (1652). *Arcana Microcosmi*[104]. Book II. London. Retrieved 2015-07-07.

- Shapiro, Robert (1987). *Origins: A Skeptic's Guide to the Creation of
  Life on Earth*. Toronto; New York: Bantam Books. ISBN 0-553-34355-
  6.

- Sheldon, Robert B. (22 September 2005). Hoover, Richard B.; Levin,
  Gilbert V.; Rozanov, Alexei Y.; Gladstone, G. Randall, eds. *Histori-
  cal Development of the Distinction between Bio- and Abiogenesis*[105]
  (PDF). Astrobiology and Planetary Missions[106]. **5906**. Bellingham, WA:
  SPIE. doi: 10.1117/12.663480[107]. ISBN 978-0-8194-5911-4. LCCN

2005284378[108]. Retrieved 2015-04-13. Proceedings of the SPIE held at San Diego, CA, 31 July–2 August 2005

- Tyndall, John (1905) [Originally published 1871; London; New York: Longmans, Green & Co.; D. Appleton and Company]. *Fragments of Science*[109]. **2** (6th ed.). New York: P.F. Collier & Sons. OCLC 726998155[110]. Retrieved 2015-06-06.
- Vartanian, Aram (1973). "Spontaneous Generation"[111]. In Wiener, Philip P. *Dictionary of the History of Ideas*. **IV**. New York: Charles Scribner's Sons. ISBN 0-684-13293-1. LCCN 72007943[112]. Retrieved 2015-06-05.
- Voet, Donald; Voet, Judith G. (2004). *Biochemistry*. **1** (3rd ed.). New York: John Wiley & Sons. ISBN 0-471-19350-X. LCCN 2003269978[113].
- Woodward, Robert J., ed. (1969). *Our Amazing World of Nature: Its Marvels & Mysteries*. Pleasantville, NY: Reader's Digest Association. ISBN 0-340-13000-8. LCCN 69010418[114].
- Yarus, Michael (2010). *Life from an RNA World: The Ancestor Within*. Cambridge, MA: Harvard University Press. ISBN 978-0-674-05075-4. LCCN 2009044011[115].
- Arrhenius, Gustaf O.; Sales, Brian C.; Mojzsis, Stephen J.; et al. (21 August 1997). "Entropy and Charge in Molecular Evolution—the Case of Phosphate"[116] (PDF). *Journal of Theoretical Biology*. Amsterdam, the Netherlands: Elsevier. **187** (4): 503–522. doi: 10.1006/jtbi.1996.0385[117]. ISSN 0022-5193[118]. PMID 9299295[119].
- Cavalier-Smith, Thomas (June 2006). "Cell evolution and Earth history: stasis and revolution"[120]. *Philosophical Transactions of the Royal Society B*. London: Royal Society. **361** (1470): 969–1006. doi: 10.1098/rstb.2006.1842[121]. ISSN 0962-8436[122]. PMC 1578732[120] ⊚. PMID 16754610[123].
- Fernando, Chrisantha T.; Rowe, Jonathan (7 July 2007). "Natural selection in chemical evolution". *Journal of Theoretical Biology*. Amsterdam, the Netherlands. **247** (1): 152–167. doi: 10.1016/j.jtbi.2007.01.028[124]. ISSN 0022-5193[118]. PMID 17399743[125].
- Gross, Michael (19 December 2016). "How life can arise from chemistry". *Current Biology*. **26** (24): R1247–R1249. doi: 10.1016/j.cub.2016.12.001[126].
- Horgan, John (February 1991). "In the Beginning.". *Scientific American*. Stuttgart: Georg von Holtzbrinck Publishing Group. **264** (2): 116–125. Bibcode: 1991SciAm.264b.116H[127]. doi: 10.1038/scientificamerican0291-116[128]. ISSN 0036-8733[129].
- Ignatov, Ignat; Mosin, Oleg V. (2013). "Modeling of Possible Processes for Origin of Life and Living Matter in Hot Mineral and Seawater with Deuterium"[130]. *Journal of Environment and Earth Science*. New York:

International Institute for Science, Technology and Education. **3** (14): 103–118. ISSN 2224-3216[131]. Retrieved 2015-06-29.

- Jortner, Joshua (October 2006). "Conditions for the emergence of life on the early Earth: summary and reflections"[132]. *Philosophical Transactions of the Royal Society B*. London: Royal Society. **361** (1474): 1877–1891. doi: 10.1098/rstb.2006.1909[133]. ISSN 0962-8436[122]. PMC 1664691[132] ⊝. PMID 17008225[134].

- Klotz, Irene (24 February 2012). "Did Life Start in a Pond, Not Oceans?"[135]. *Discovery News*. Silver Spring, MD: Discovery Communications. Retrieved 2015-06-29.

- NASA Astrobiology Institute: Harrison, T. Mark; McKeegan, Kevin D.; Mojzsis, Stephen J. "Earth's Early Environment and Life: When did Earth become suitable for habitation?"[136]. Archived from the original[137] on 2012-02-17. Retrieved 2015-06-30.

- NASA Specialized Center of Research and Training in Exobiology: Arrhenius, Gustaf O. (11 September 2002). "Arrhenius"[138]. Archived from the original[139] on 2007-12-21. Retrieved 2015-06-30.

- "The physico-chemical basis of life"[140]. *What is Life*. Spring Valley, CA: Lukas K. Buehler. Retrieved 27 October 2005.

- Pitsch, Stefan; Krishnamurthy, Ramanarayanan; Arrhenius, Gustaf O. (6 September 2000). "Concentration of Simple Aldehydes by Sulfite-Containing Double-Layer Hydroxide Minerals: Implications for Biopoesis". *Helvetica Chimica Acta*. Hoboken, NJ: John Wiley & Sons. **83** (9): 2398–2411. doi: 10.1002/1522-2675(20000906)83:9<2398::AID-HLCA2398>3.0.CO;2-5[141]. ISSN 0018-019X[142]. PMID 11543578[143].

- Pons, Marie-Laure; Quitté, Ghylaine; Fujii, Toshiyuki; et al. (25 October 2011). "Early Archean Serpentine Mud Volcanoes at Isua, Greenland, as a Niche for Early Life"[144]. *Proc. Natl. Acad. Sci. U.S.A.* Washington, D.C.: National Academy of Sciences. **108** (43): 17639–17643. Bibcode: 2011PNAS..10817639P[145]. doi: 10.1073/pnas.1108061108[146]. ISSN 0027-8424[147]. PMC 3203773[144] ⊝. PMID 22006301[148].

- Russell, Michael J.; Hall, A. J.; Cairns-Smith, Alexander Graham; et al. (10 November 1988). "Submarine hot springs and the origin of life". *Nature*. London: Nature Publishing Group. **336** (6195): 117. Bibcode: 1988Natur.336..117R[149]. doi: 10.1038/336117a0[150]. ISSN 0028-0836[151]. More than one of |pages= and |page= specified (help)

- Shock, Everett L. (25 October 1997). "High-temperature life without photosynthesis as a model for Mars"[152] (PDF). *Journal of Geophysical Research*. Washington, D.C.: American Geophysical Union. **102** (E10): 23687–23694. Bibcode: 1997JGR...10223687S[153]. doi: 10.1029/97je01087[154]. ISSN 0148-0227[155].

External links

<indicator name="spoken-icon"> ◉⟩⟩ </indicator>

シ ㋐ ㋬ λ ㋷ ㋞ ホ 維 ㋦	Look up *abiogenesis* in Wiktionary, the free dictionary.

- "Exploring Life's Origins: A Virtual Exhibit"[156]. *Exploring Life's Origins: A Virtual Exhibit*. Arlington County, VA: National Science Foundation. Retrieved 2015-07-02.
- "The Geochemical Origins of Life by Michael J. Russell & Allan J. Hall"[157]. Glasgow, Scotland: University of Glasgow. 13 December 2008. Retrieved 2015-07-02.
- Malory, Marcia. "How life began on Earth"[158]. *Earth Facts*. Retrieved 2015-07-02.
- The Origins of Life[159], BBC Radio 4 discussion with Richard Dawkins, Richard Corfield & Linda Partridge (*In Our Time*, Sep. 23, 2004)

Early geophysical conditions on Earth

Timeline of the evolutionary history of life

Life timeline

θ —
⸻
500—
1000—
1500—
2000—
2500—
3000—
3500—
4000—
4500—

Axis scale: million years

Also see: *Human timeline* and *Nature timeline*

This **timeline of the evolutionary history of life** represents the current scientific theory outlining the major events during the development of life on planet Earth. In biology, evolution is any change across successive generations in the heritable characteristics of biological populations. Evolutionary processes give rise to diversity at every level of biological organization, from kingdoms to species, and individual organisms and molecules, such as DNA and proteins. The similarities between all present day organisms indicate the presence of a common ancestor from which all known species, living and extinct, have diverged through the process of evolution. More than 99 percent of all species, amounting to over five billion species, that ever lived on Earth are estimated to be extinct. Estimates on the number of Earth's current species range from 10 million to 14 million, of which about 1.2 million have been documented and over 86 percent have not yet been described. However, a May 2016 scientific report estimates that 1 trillion species are currently on Earth, with only one-thousandth of one percent described.

While the dates given in this article are estimates based on scientific evidence, there has been controversy between more traditional views of increased biodiversity through a cone of diversity with the passing of time and the view that the basic pattern on Earth has been one of annihilation and diversification and that in certain past times, such as the Cambrian explosion, there was great diversity.[160]

Extinction

Species go extinct constantly as environments change, as organisms compete for environmental niches, and as genetic mutation leads to the rise of new species from older ones. Occasionally biodiversity on Earth takes a hit in the form of a mass extinction in which the extinction rate is much higher than usual.[161] A large extinction-event often represents an accumulation of smaller extinction- events that take place in a relatively brief period of time.[162]

The first known mass extinction in earth's history was the Great Oxygenation Event 2.4 billion years ago. That event led to the loss of most of the planet's obligate anaerobes. Researchers have identified five major extinction events in earth's history since:[163]

- End of the Ordovician: 440 million years ago, 86% of all species lost, including graptolites

Figure 12: *Visual representation of the history of life on Earth as a spiral*

- Late Devonian: 375 million years ago, 75% of species lost, including most trilobites
- End of the Permian, "The Great Dying": 251 million years ago, 96% of species lost, including tabulate corals, and most extant trees and synapsids
- End of the Triassic: 200 million years ago, 80% of species lost, including all of the conodonts
- End of the Cretaceous: 66 million years ago, 76% of species lost, including all of the ammonites, mosasaurs, ichthyosaurs, plesiosaurs, pterosaurs, and nonavian dinosaurs

(Dates and percentages represent estimates.)

Smaller extinction-events have occurred in the periods between these larger catastrophes, with some standing at the delineation points of the periods and epochs recognized by scientists in geologic time. The Holocene extinction event is currently under way.

Factors in mass extinctions include continental drift, changes in atmospheric and marine chemistry, volcanism and other aspects of mountain formation, changes in glaciation, changes in sea level, and impact events.

Figure 13: *Moon*

Detailed timeline

In this timeline, **Ma** (for *megaannum*) means "million years ago," **ka** (for *kiloannum*) means "thousand years ago," and **ya** means "years ago."

Hadean Eon

4000 Ma and earlier.

Date	Event
4600 Ma	The planet Earth forms from the accretion disc revolving around the young Sun with organic compounds (complex organic molecules) necessary for life having perhaps formed in the protoplanetary disk of cosmic dust grains surrounding it before the formation of the Earth.
4500 Ma	According to the giant impact hypothesis, the Moon was formed when the planet Earth and the hypothesized planet Theia collided, sending a very large number of moonlets into orbit around the young Earth which eventually coalesced to form the Moon. The gravitational pull of the new Moon stabilised the Earth's fluctuating axis of rotation and set up the conditions in which abiogenesis occurred.
4404 Ma	First appearance of liquid water on Earth.
4280 Ma	Earliest possible appearance of life on Earth.

Figure 14: *Fragment of the Acasta Gneiss exhibited at the Museum of Natural History in Vienna*

Archean Eon

4000 Ma – 2500 Ma

Date	Event
4000 Ma	Formation of a greenstone belt of the Acasta Gneiss of the Slave craton in Northwest Territories, Canada, the oldest rock belt in the world.
4100–3800 Ma	Late Heavy Bombardment (LHB): extended barrage of impact events upon the inner planets by meteoroids. Thermal flux from widespread hydrothermal activity during the LHB may have been conducive to abiogenesis and life's early diversification. "Remains of biotic life" were found in 4.1 billion-year-old rocks in Western Australia. According to one of the researchers, "If life arose relatively quickly on Earth ... then it could be common in the universe."Wikipedia:Writing better articles#Stay on topic This is when life most likely arose.
3900–2500 Ma	Cells resembling prokaryotes appear. These first organisms are chemoautotrophs: they use carbon dioxide as a carbon source and oxidize inorganic materials to extract energy. Later, prokaryotes evolve glycolysis, a set of chemical reactions that free the energy of organic molecules such as glucose and store it in the chemical bonds of ATP. Glycolysis (and ATP) continue to be used in almost all organisms, unchanged, to this day.

3800 Ma	Formation of a greenstone belt of the Isua complex of the western Greenland region, whose rocks show an isotope frequency suggestive of the presence of life. The earliest evidences for life on Earth are 3.8 billion-year-old biogenic hematite in a banded iron formation of the Nuvvuagittuq Greenstone Belt in Canada,[164] graphite in 3.7 billion-year-old metasedimentary rocks discovered in western Greenland and microbial mat fossils found in 3.48 billion-year-old sandstone discovered in Western Australia.
3500 Ma	Lifetime of the last universal common ancestor (LUCA); the split between bacteria and archaea occurs. Bacteria develop primitive forms of photosynthesis which at first did not produce oxygen. These organisms generated Adenosine triphosphate by exploiting a proton gradient, a mechanism still used in virtually all organisms.[165]
3200 Ma	Diversification and expansion of acritarchs.
3000 Ma	Photosynthesizing cyanobacteria evolved; they used water as a reducing agent, thereby producing oxygen as a waste product. The oxygen initially oxidizes dissolved iron in the oceans, creating iron ore. The oxygen concentration in the atmosphere slowly rose, acting as a poison for many bacteria and eventually triggering the Great Oxygenation Event. The Moon, still very close to Earth, caused tides 1,000 feet (305 m) high.Wikipedia:Citation needed The Earth was continually wracked by hurricane-force winds. These extreme mixing influences are thought to have stimulated evolutionary processes.Wikipedia:Citation needed
2800 Ma	Oldest evidence for microbial life on land in the form of organic matter-rich paleosols, ephemeral ponds and alluvial sequences, some of them bearing microfossils.

Proterozoic Eon

2500 Ma – 542 Ma. Contains the Palaeoproterozoic, Mesoproterozoic and Neoproterozoic eras.

Date	Event
2500 Ma	Great Oxygenation Event led by cyanobacteria's oxygenic photosynthesis. Commencement of plate tectonics with old marine crust dense enough to subduct.
By 1850 Ma	Eukaryotic cells appear. Eukaryotes contain membrane-bound organelles with diverse functions, probably derived from prokaryotes engulfing each other via phagocytosis. (See Symbiogenesis and Endosymbiont). Bacterial viruses (bacteriophage) emerge before, or soon after, the divergence of the prokaryotic and eukaryotic lineages. The appearance of red beds show that an oxidising atmosphere had been produced. Incentives now favoured the spread of eukaryotic life.
1400 Ma	Great increase in stromatolite diversity.
1300 Ma	Earliest land fungi
By 1200 Ma	Meiosis and sexual reproduction are present in single-celled eukaryotes, and possibly in the common ancestor of all eukaryotes. Sex may even have arisen earlier in the RNA world. Sexual reproduction first appears in the fossil records; it may have increased the rate of evolution.
1 bya	The first non-marine eukaryotes move onto land. They were photosynthetic and multicellular, indicating that plants evolved much earlier than originally thought.

750 Ma	First protozoa (ex: *Melanocyrillium*)
850–630 Ma	A global glaciation may have occurred. Opinion is divided on whether it increased or decreased biodiversity or the rate of evolution. It is believed that this was due to evolution of first land plants, which increased the amount of oxygen and lowered the number of carbon dioxide in the atmosphere.
600 Ma	The accumulation of atmospheric oxygen allows the formation of an ozone layer. Prior to this, land-based life would probably have required other chemicals to attenuate ultraviolet radiation enough to permit colonisation of the land.
580–542 Ma	The Ediacara biota represent the first large, complex aquatic multicellular organisms — although their affinities remain a subject of debate.
580–500 Ma	Most modern phyla of animals begin to appear in the fossil record during the Cambrian explosion.
550 Ma	First fossil evidence for Ctenophora (comb jellies), Porifera (sponges), Anthozoa (corals and sea anemones)

Phanerozoic Eon

542 Ma – present

The Phanerozoic Eon, literally the "period of well-displayed life," marks the appearance in the fossil record of abundant, shell-forming and/or trace-making organisms. It is subdivided into three eras, the Paleozoic, Mesozoic and Cenozoic, which are divided by major mass extinctions.

Palaeozoic Era

542 Ma – 251.0 Ma and contains the Cambrian, Ordovician, Silurian, Devonian, Carboniferous and Permian periods.

Date	Event
535 Ma	Major diversification of living things in the oceans: chordates, arthropods (e.g. trilobites, crustaceans), echinoderms, molluscs, brachiopods, foraminifers and radiolarians, etc.
530 Ma	The first known footprints on land date to 530 Ma.
525 Ma	Earliest graptolites
510 Ma	First cephalopods (nautiloids) and chitons
505 Ma	Fossilization of the Burgess Shale
485 Ma	First vertebrates with true bones (jawless fishes)
450 Ma	First complete conodonts and echinoids appear

440 Ma	First agnathan fishes: Heterostraci, Galeaspida, and Pituriaspida
420 Ma	Earliest ray-finned fishes, trigonotarbid arachnids, and land scorpions
410 Ma	First signs of teeth in fish. Earliest Nautilida, lycophytes, and trimerophytes.
395 Ma	First lichens, stoneworts. Earliest harvestmen, mites, hexapods (springtails) and ammonoids. The first known tetrapod tracks on land.
363 Ma	By the start of the Carboniferous Period, the Earth begins to resemble its present state. Insects roamed the land and would soon take to the skies; sharks swam the oceans as top predators, and vegetation covered the land, with seed-bearing plants and forests soon to flourish. Four-limbed tetrapods gradually gain adaptations which will help them occupy a terrestrial life-habit.
360 Ma	First crabs and ferns. Land flora dominated by seed ferns.
350 Ma	First large sharks, ratfishes, and hagfish
340 Ma	Diversification of amphibians
330 Ma	First amniote vertebrates (*Paleothyris*)
320 Ma	Synapsids (precursors to mammals) separate from sauropsids (reptiles) in late Carboniferous.
305 Ma	Earliest diapsid reptiles (e.g. *Petrolacosaurus*)
280 Ma	Earliest beetles, seed plants and conifers diversify while lepidodendrids and sphenopsids decrease. Terrestrial temnospondyl amphibians and pelycosaurs (e.g. *Dimetrodon*) diversify in species.
275 Ma	Therapsid synapsids separate from pelycosaur synapsids
251.4 Ma	The Permian–Triassic extinction event eliminates over 90-95% of marine species. Terrestrial organisms were not as seriously affected as the marine biota. This "clearing of the slate" may have led to an ensuing diversification, but life on land took 30 million years to completely recover.

Mesozoic Era

From 251.4 Ma to 66 Ma and containing the Triassic, Jurassic and Cretaceous periods.

Date	Event
	The Mesozoic Marine Revolution begins: increasingly well adapted and diverse predatorsWikipedia:Manual of Style/Words to watch#Unsupported attributions pressurize sessile marine groups; the "balance of power" in the oceans shifts dramatically as some groups of preyWikipedia:Manual of Style/Words to watch#Unsupported attributions adapt more rapidly and effectively than othersWikipedia:Manual of Style/Words to watch#Unsupported attributions.

248 Ma	Sturgeon and paddlefish (Acipenseridae) first appear.
245 Ma	Earliest ichthyosaurs
240 Ma	Increase in diversity of gomphodont cynodonts and rhynchosaurs
225 Ma	Earliest dinosaurs (prosauropods), first cardiid bivalves, diversity in cycads, bennetti-taleans, and conifers. First teleost fishes. First mammals (*Adelobasileus*).
220 Ma	Seed-producing Gymnosperm forests dominate the land; herbivores grow to huge sizes to accommodate the large guts necessary to digest the nutrient-poor plants.Wikipedia:Citation needed First flies and turtles (*Odontochelys*). First coelophysoid dinosaurs.
205 Ma	the Massive extinction of Triassic/Jurassic, that wiped out most of the group of pseudosuchians and was given the opportunity of dinosaurs including the Apatosaurus, Tyrannosaurus, Perrotasaurus, and Stegosaurus to enter its golden age.
200 Ma	The first accepted evidence for viruses that infect eukaryotic cells (at least, the group Geminiviridae) existed. Viruses are still poorly understood and may have arisen before "life" itself, or may be a more recent phenomenon. Major extinctions in terrestrial vertebrates and large amphibians. Earliest examples of armoured dinosaurs
195 Ma	First pterosaurs with specialized feeding (*Dorygnathus*). First sauropod dinosaurs. Diversification in small, ornithischian dinosaurs: heterodontosaurids, fabrosaurids, and scelidosaurids.
190 Ma	Pliosauroids appear in the fossil record. First lepidopteran insects (*Archaeolepis*), hermit crabs, modern starfish, irregular echinoids, corbulid bivalves, and tubulipore bryozoans. Extensive development of sponge reefs.
176 Ma	First members of the Stegosauria group of dinosaurs
170 Ma	Earliest salamanders, newts, cryptoclidids, elasmosaurid plesiosaurs, and cladotherian mammals. Sauropod dinosaurs diversify.
165 Ma	First rays and glycymeridid bivalves
163 Ma	Pterodactyloid pterosaurs first appear
161 Ma	Ceratopsian dinosaurs appear in the fossil record (*Yinlong*) and the oldest known Eutherian Mammal appear in the fossil record: *Juramaia*.
160 Ma	Multituberculate mammals (genus *Rugosodon*) appear in eastern China
155 Ma	First blood-sucking insects (ceratopogonids), rudist bivalves, and cheilostome bryozoans. *Archaeopteryx*, a possible ancestor to the birds, appears in the fossil record, along with triconodontid and symmetrodont mammals. Diversity in stegosaurian and theropod dinosaurs.
130 Ma	The rise of the angiosperms: Some of these flowering plants bear structures that attract insects and other animals to spread pollen;other angiosperms were pollinated by wind or water. This innovation causes a major burst of animal evolution through coevolution. First freshwater pelomedusid turtles.
120 Ma	Oldest fossils of heterokonts, including both marine diatoms and silicoflagellates
115 Ma	First monotreme mammals

110 Ma	First hesperornithes, toothed diving birds. Earliest limopsid, verticordiid, and thyasirid bivalves.
106 Ma	*Spinosaurus*, the largest theropod dinosaur, appears in the fossil record
100 Ma	Earliest bees
90 Ma	Extinction of ichthyosaurs. Earliest snakes and nuculanid bivalves. Large diversification in angiosperms: magnoliids, rosids, hamamelidids, monocots, and ginger. Earliest examples of ticks. Probable origins of placental mammals (earliest undisputed fossil evidence is 66 Ma).
80 Ma	First ants
70 Ma	Multituberculate mammals increase in diversity. First yoldiid bivalves.
68 Ma	*Tyrannosaurus*, the largest terrestrial predator of what is now western North America appears in the fossil record. First species of *Triceratops*.

Cenozoic Era

66 Ma – present

Hominin timeline

Figure 15: *The cyanobacterial-algal mat, salty lake on the White Sea seaside*

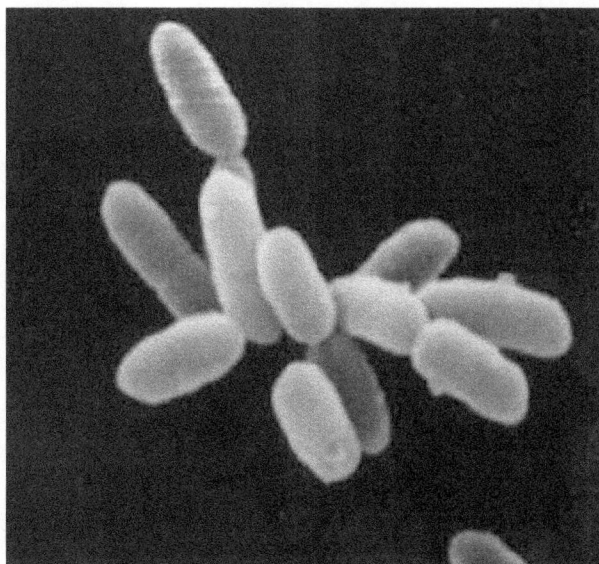

Figure 16: *Halobacterium sp. strain NRC-1*

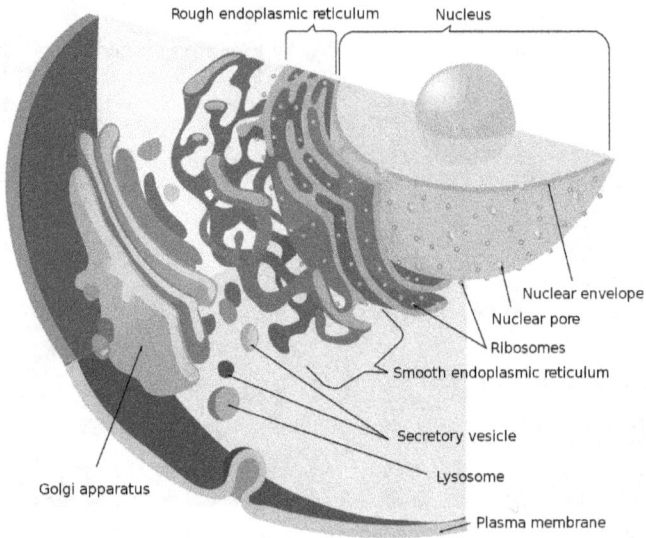

Rough endoplasmic reticulum   Nucleus

Nuclear envelope
Nuclear pore
Ribosomes
Smooth endoplasmic reticulum

Secretory vesicle
Lysosome

Golgi apparatus

Plasma membrane

Figure 17: *Detail of the eukaryote endomembrane system and its components*

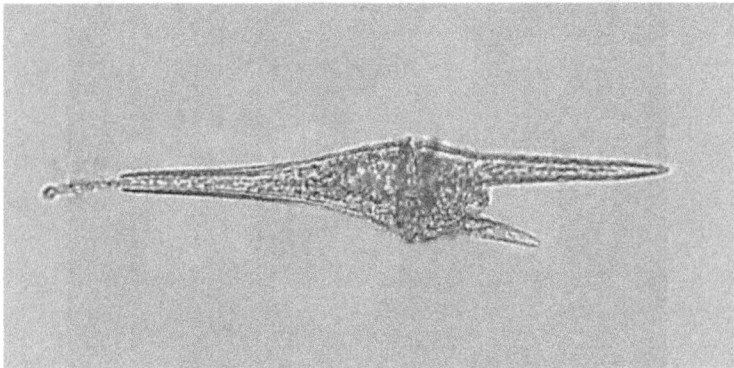

Figure 18: *Dinoflagellate Ceratium furca*

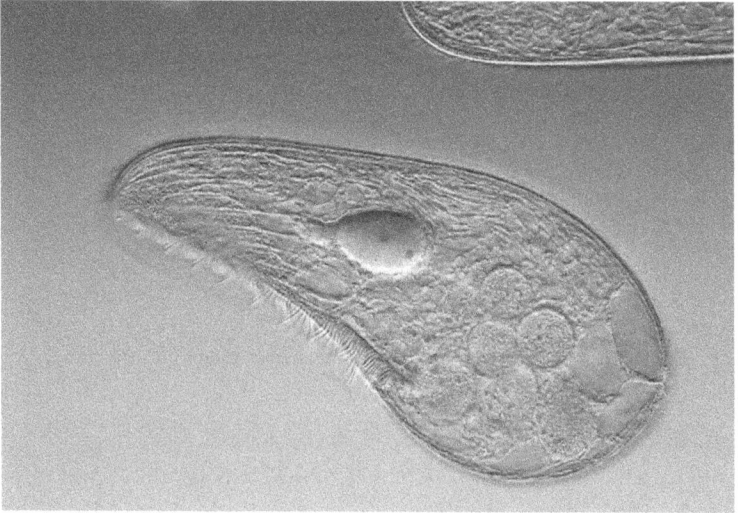

Figure 19: *Blepharisma japonicum, a free-living ciliated protozoan*

Figure 20: *Dickinsonia costata, an iconic Ediacaran organism, displays the characteristic quilted appearance of Ediacaran enigmata.*

Figure 21: *Utatsusaurus is the earliest-known form of an ichthyopterygian.*

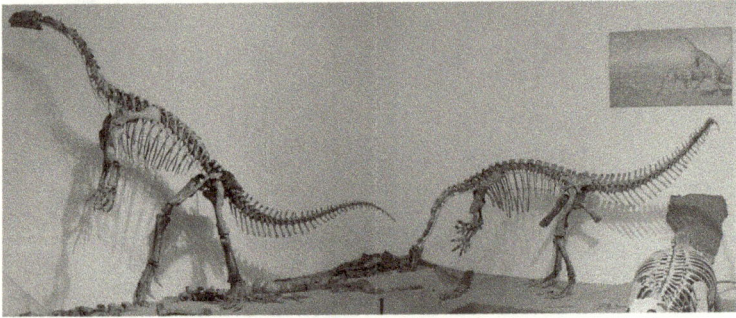

Figure 22: *Plateosaurus engelhardti*

Axis scale: million years

Also see: *Life timeline* and *Nature timeline*

Date	Event
66 Ma	The Cretaceous–Paleogene extinction event eradicates about half of all animal species, including mosasaurs, pterosaurs, plesiosaurs, ammonites, belemnites, rudist and inoceramid bivalves, most planktic foraminifers, and all of the dinosaurs excluding the birds.
From 66 Ma	Rapid dominance of conifers and ginkgos in high latitudes, along with mammals becoming the dominant species. First psammobiid bivalves. Earliest rodents. Rapid diversification in ants.
63 Ma	Evolution of the creodonts, an important group of meat-eating (carnivorous) mammals

60 Ma	Diversification of large, flightless birds. Earliest true primates, along with the first semelid bivalves, edentate, carnivoran and lipotyphlan mammals, and owls. The ancestors of the carnivorous mammals (miacids) were alive.
56 Ma	*Gastornis*, a large flightless bird, appears in the fossil record
55 Ma	Modern bird groups diversify (first song birds, parrots, loons, swifts, woodpeckers), first whale (*Himalayacetus*), earliest lagomorphs, armadillos, appearance of sirenian, proboscidean, perissodactyl and artiodactyl mammals in the fossil record. Angiosperms diversify. The ancestor (according to theory) of the species in the genus *Carcharodon*, the early mako shark *Isurus hastalis*, is alive.
52 Ma	First bats appear (*Onychonycteris*)
50 Ma	Peak diversity of dinoflagellates and nannofossils, increase in diversity of anomalodesmatan and heteroconch bivalves, brontotheres, tapirs, rhinoceroses, and camels appear in the fossil record, diversification of primates
40 Ma	Modern-type butterflies and moths appear. Extinction of *Gastornis*. *Basilosaurus*, one of the first of the giant whales, appeared in the fossil record.
37 Ma	First nimravid ("false saber-toothed cats") carnivores — these species are unrelated to modern-type felines
35 Ma	Grasses diversify from among the monocot angiosperms; grasslands begin to expand. Slight increase in diversity of cold-tolerant ostracods and foraminifers, along with major extinctions of gastropods, reptiles, amphibians, and multituberculate mammals. Many modern mammal groups begin to appear: first glyptodonts, ground sloths, canids, peccaries, and the first eagles and hawks. Diversity in toothed and baleen whales.
33 Ma	Evolution of the thylacinid marsupials (*Badjcinus*)
30 Ma	First balanids and eucalypts, extinction of embrithopod and brontothere mammals, earliest pigs and cats
28 Ma	*Paraceratherium* appears in the fossil record, the largest terrestrial mammal that ever lived
25 Ma	*Pelagornis sandersi* appears in the fossil record, the largest flying bird that ever lived
25 Ma	First deer
20 Ma	First giraffes, hyenas, bears and giant anteaters, increase in bird diversity
15 Ma	Genus *Mammut* appears in the fossil record, first bovids and kangaroos, diversity in Australian megafauna
10 Ma	Grasslands and savannas are established, diversity in insects, especially ants and termites, horses increase in body size and develop high-crowned teeth, major diversification in grassland mammals and snakes
9.5 Ma	The Great American Interchange, where various land and freshwater faunas migrated between North and South America. Armadillos, opossums, hummingbirds Phorusrhacids, Ground Sloths, Glyptodonts, and Meridiungulates traveled to North America, while horses, tapirs, saber-toothed cats, Jaguars, Bears, Coaties, Ferrets, Otters, Skunks and deer entered South America.
6.5 Ma	First hominins (*Sahelanthropus*)

6 Ma	Australopithecines diversify (*Orrorin, Ardipithecus*)
5 Ma	First tree sloths and hippopotami, diversification of grazing herbivores like zebras and elephants, large carnivorous mammals like lions and the genus *Canis*, burrowing rodents, kangaroos, birds, and small carnivores, vultures increase in size, decrease in the number of perissodactyl mammals. Extinction of nimravid carnivores.
4.8 Ma	Mammoths appear in the fossil record
4 Ma	Evolution of *Australopithecus*, *Stupendemys* appears in the fossil record as the largest freshwater turtle, first modern elephants, giraffes, zebras, lions, rhinoceros and gazelles appear in the fossil record
2.7 Ma	Evolution of *Paranthropus*
2.5 Ma	The earliest species of *Smilodon* evolve
2 Ma	First members of the genus *Homo*, Homo Habilis, appear in the fossil record. Diversification of conifers in high latitudes. The eventual ancestor of cattle, aurochs (*Bos primigenus*), evolves in India.
1.7 Ma	Extinction of australopithecines
1.2 Ma	Evolution of *Homo antecessor*. The last members of *Paranthropus* die out.
800 Ka	Short-faced bears (*Arctodus simus*) become abundant in North America
600 ka	Evolution of *Homo heidelbergensis*
350 ka	Evolution of Neanderthals
300 ka	*Gigantopithecus*, a giant relative of the orangutan from Asia dies out
250 ka	Anatomically modern humans appear in Africa. Around 50,000 years before present they start colonising the other continents, replacing the Neanderthals in Europe and other hominins in Asia.
40 ka	The last of the giant monitor lizards (Varanus priscus) die out
30 ka	Extinction of Neanderthals, first domestic dogs
15 ka	The last woolly rhinoceros (*Coelodonta antiquitatis*) are believed to have gone extinct
11 ka	Short-faced bears vanish from North America, with the last giant ground sloths dying out. All Equidae become extinct in North America.
10 ka	The Holocene epoch starts 10,000 years ago after the Late Glacial Maximum. The last mainland species of woolly mammoth (*Mammuthus primigenus*) die out, as does the last *Smilodon* species.
8 ka	The Giant Lemur died out

Figure 23: *Cycas circinalis*

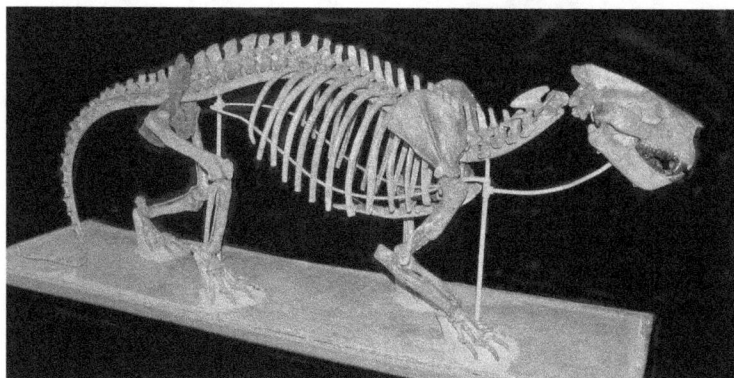

Figure 24: *Mount of oxyaenid Patriofelis from
the American Museum of Natural History*

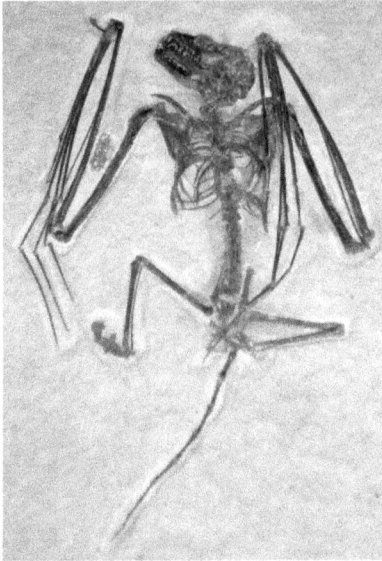

Figure 25: *The bat Icaronycteris appeared 52.2 million years ago*

Figure 26: *Grass flowers*

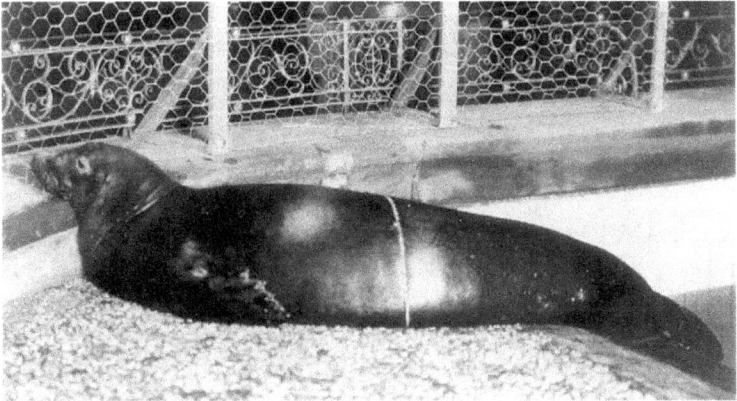

Figure 27: *Caribbean monk seal*

Figure 28: *Illustration of a Baiji, declared functionally extinct by the Baiji.org Foundation in 2006.*

Historical extinctions

Date	Event
6000 ya (c. 4000 BC)	Small populations of American mastodon die off in places like Utah and Michigan
4500 ya (c. 2500 BC)	The last members of a dwarf race of woolly mammoths vanish from Wrangel Island near Alaska
c. 600 ya (c. 1400)	The moa and its predator, Haast's eagle, die out in New Zealand
391 ya (1627)	The last recorded wild aurochs die out
330 ya (1688)	The dodo goes extinct

250 ya (1768)	The Steller's sea cow goes extinct
135 ya (1883)	The quagga, a subspecies of zebra, goes extinct
104 ya (1914)	Martha, last known passenger pigeon, dies
82 ya (1936)	The thylacine goes extinct in a Tasmanian zoo, the last member of the family Thylacinidae
66 ya (1952)	The Caribbean monk seal goes extinct
10 ya (2008)	The baiji, the Yangtze river dolphin, becomes functionally extinct, according to the IUCN Red List
7 ya (2011)	The western black rhinoceros is declared extinct

Bibliography

<templatestyles src="Template:Refbegin/styles.css" />

- Barton, Nicholas H.; Briggs, Derek E.G.; Eisen, Jonathan A.; Goldstein, David B.; Patel, Nipam H. (2007). *Evolution*. Cold Spring Harbor, NY: Cold Spring Harbor Laboratory Press. ISBN 978-0-87969-684-9. LCCN 2007010767[166]. OCLC 86090399[167].
- Bernstein, Harris; Bernstein, Carol; Michod, Richard E. (2012). "DNA Repair as the Primary Adaptive Function of Sex in Bacteria and Eukaryotes"[168]. In Kimura, Sakura; Shimizu, Sora. *DNA Repair: New Research*. Hauppauge, NY: Nova Science Publishers. ISBN 978-1-62100-808-8. LCCN 2011038504[169]. OCLC 828424701[170].
- Bjornerud, Marcia (2005). *Reading the Rocks: The Autobiography of the Earth*. Cambridge, MA: Westview Press. ISBN 0-8133-4249X. LCCN 2004022738[171]. OCLC 56672295[172].
- Kirschvink, Joseph L. (1992). "Late Proterozoic Low-Latitude Global Glaciation: the Snowball Earth"[173] (PDF). In Schopf, J. William; Klein, Cornelis. *The Proterozoic Biosphere: A Multidisciplinary Study*. Cambridge; New York: Cambridge University Press. ISBN 0-521-36615-1. LCCN 91015085[174]. OCLC 23583672[175].
- McKinney, Michael L. (1997). "How do rare species avoid extinction? A paleontological view". In Kunin, William E.; Gaston, Kevin J. *The Biology of Rarity: Causes and consequences of rare—common differences* (1st ed.). London; New York: Chapman & Hall. ISBN 0-412-63380-9. LCCN 96071014[176]. OCLC 36442106[177].
- Miller, G. Tyler; Spoolman, Scott E. (2012). *Environmental Science* (14th ed.). Belmont, CA: Brooks/Cole. ISBN 978-1-111-98893-7. LCCN 2011934330[178]. OCLC 741539226[179].
- Stearns, Beverly Peterson; Stearns, Stephen C. (1999). *Watching, from the Edge of Extinction*. New Haven, CT: Yale University Press. ISBN 0-300-07606-1. LCCN 98034087[180]. OCLC 47011675[181].

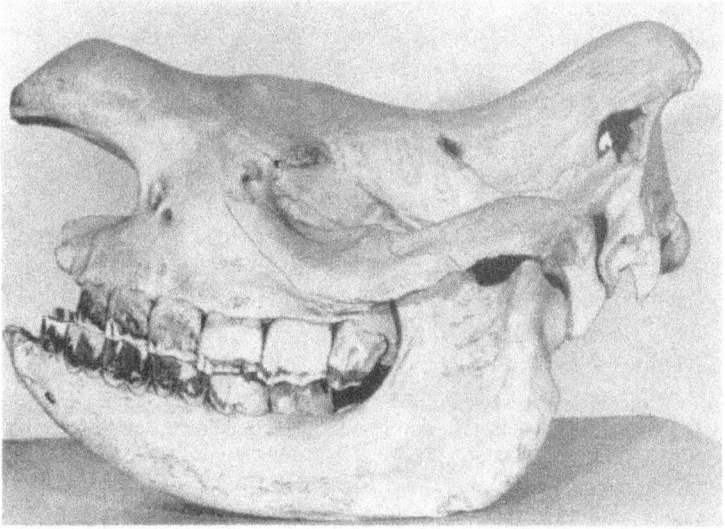

Figure 29: *Western black rhinoceros, holotype specimen of a female shot in 1911*

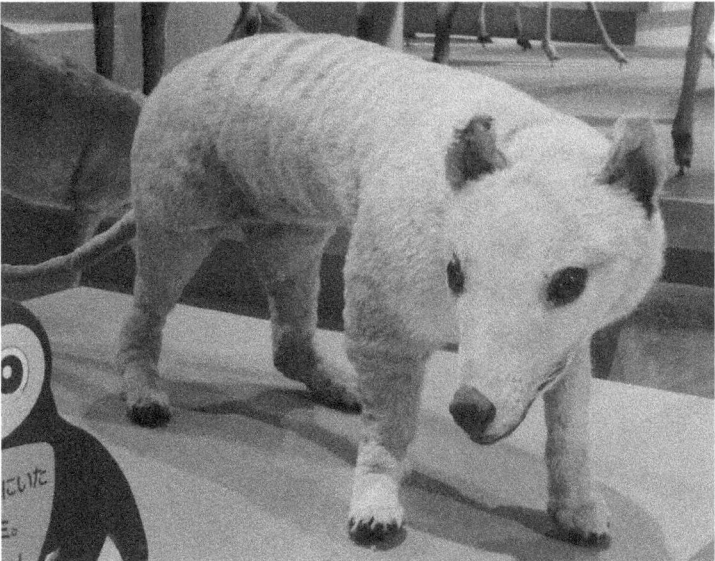

Figure 30: *Thylacine shot in 1936*

Further reading

- Dawkins, Richard (2004). *The Ancestor's Tale: A Pilgrimage to the Dawn of Life*. Boston: Houghton Mifflin Company. ISBN 0-618-00583-8. LCCN 2004059864[182]. OCLC 56617123[183].

External links

- "Understanding Evolution: your one-stop resource for information on evolution"[184]. University of California, Berkeley. Retrieved 2015-03-18.
- "Life on Earth"[185]. *Tree of Life Web Project*. University of Arizona. January 1, 1997. Retrieved 2015-03-18. Explore complete phylogenetic tree interactively
- Brandt, Niel. "Evolutionary and Geological Timelines"[186]. *TalkOrigins Archive*. Houston, TX: The TalkOrigins Foundation, Inc. Retrieved 2015-03-18.
- "Palaeos: Life Through Deep Time"[187]. *Palaeos*. Retrieved 2015-03-18.
- Kyrk, John. "Evolution"[188] (SWF). *Cell Biology Animation*. Retrieved 2015-03-18. Interactive timeline from Big Bang to present
- "Plant Evolution"[189]. *Plant and Animal Evolution*. University of Waikato. Retrieved 2015-03-18. Sequence of Plant Evolution
- "The History of Animal Evolution"[190]. *Plant and Animal Evolution*. University of Waikato. Retrieved 2015-03-18. Sequence of Animal Evolution
- Yeo, Dannel; Drage, Thomas (2006). "History of Life on Earth"[191]. Archived from the original[192] on 2015-03-15. Retrieved 2015-03-19.
- *Exploring Time*[193]. *The Science Channel*. 2007. Retrieved 2015-03-19.
- Roberts, Ben. "Plant evolution timeline"[194]. University of Cambridge. Retrieved 2015-03-19.
- Art of the Nature Timelines on Wikipedia[195]

Earliest known life forms

The **earliest known life forms** on Earth are putative fossilized microorganisms found in hydrothermal vent precipitates. The earliest time that life forms first appeared on Earth is unknown. They may have lived earlier than 3.77 billion years ago, possibly as early as 4.28 billion years ago, not long after the oceans formed 4.41 billion years ago, and not long after the formation of the Earth 4.54 billion years ago. The earliest *direct* evidence of life on Earth are fossils of microorganisms permineralized in 3.465-billion-year-old Australian Apex chert rocks.

Overview

A life form, or lifeform, is an entity or being that is living. Currently, Earth remains the only place in the universe known to harbor life forms.

More than 99% of all species of life forms, amounting to over five billion species, that ever lived on Earth are estimated to be extinct.

Some estimates on the number of Earth's current species of life forms range from 10 million to 14 million, of which about 1.2 million have been documented and over 86 percent have not yet been described. However, a May 2016 scientific report estimates that 1 trillion species are currently on Earth, with only one-thousandth of one percent described. The total number of DNA base pairs on Earth is estimated at 5.0×10^{37} with a weight of 50 billion tonnes. In comparison, the total mass of the biosphere has been estimated to be as much as 4 trillion tons of carbon. In July 2016, scientists reported identifying a set of 355 genes from the Last Universal Common Ancestor (LUCA) of all organisms living on Earth.

The Earth's biosphere includes soil, hydrothermal vents, rock up to 19 km (12 mi) or deeper underground, the deepest parts of the ocean, and at least 64 km (40 mi) high into the atmosphere. Under certain test conditions, life forms have been observed to thrive in the near-weightlessness of space and to survive in the vacuum of outer space. Life forms appear to thrive in the Mariana Trench, the deepest spot in the Earth's oceans, reaching a depth of 11,034 m (36,201 ft; 6.856 mi). Other researchers reported related studies that life forms thrive inside rocks up to 580 m (1,900 ft; 0.36 mi) below the sea floor under 2,590 m (8,500 ft; 1.61 mi) of ocean off the coast of the northwestern United States, as well as 2,400 m (7,900 ft; 1.5 mi) beneath the seabed off Japan. In August 2014, scientists confirmed the existence of life forms living 800 m (2,600 ft; 0.50 mi) below the ice of Antarctica.

According to one researcher, "You can find microbes everywhere — they're extremely adaptable to conditions, and survive wherever they are."

Figure 31: *Evidence of possibly the oldest forms of life on Earth has been found in hydrothermal vent precipitates.*

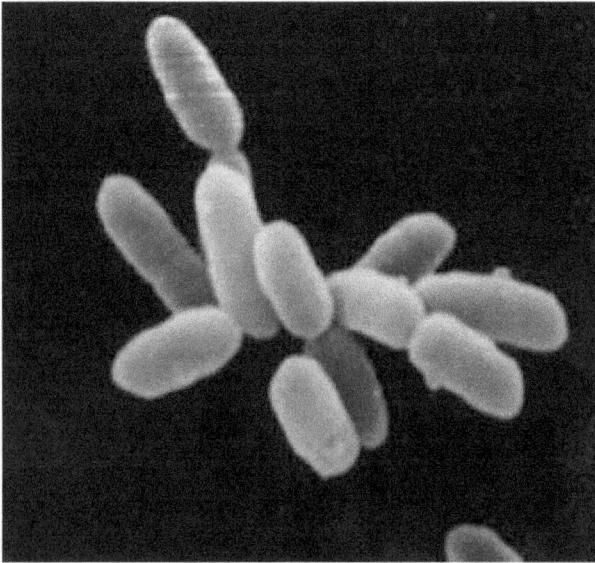

Figure 32: *Archaea, prokaroytic microbes, were first found in extreme environments, such as hydrothermal vents.*

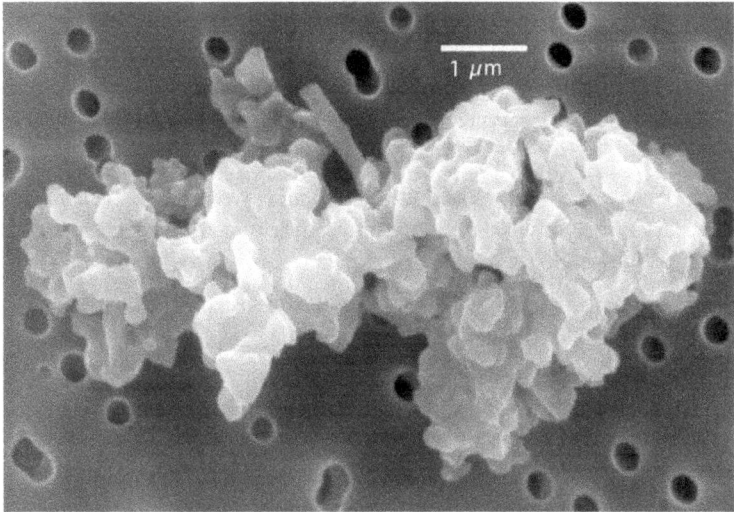

Figure 33: *Studies suggest that life on Earth may have come from biological matter carried by space dust or meteorites.*

Earliest life forms

Fossil evidence informs most studies of the origin of life. The age of the Earth is about 4.54 billion years; the earliest undisputed evidence of life on Earth dates from at least 3.5 billion years ago. There is evidence that life began much earlier.

In 2017, fossilized microorganisms, or microfossils, were announced to have been discovered in hydrothermal vent precipitates in the Nuvvuagittuq Belt of Quebec, Canada that may be as old as 4.28 billion years old, the oldest record of life on Earth, suggesting "an almost instantaneous emergence of life" (in a geological time-scale sense), after ocean formation 4.41 billion years ago, and not long after the formation of the Earth 4.54 billion years ago.

"Remains of life" have been found in 4.1 billion-year-old rocks in Western Australia.

Evidence of biogenic graphite, and possibly stromatolites, was discovered in 3.7 billion-year-old metasedimentary rocks in southwestern Greenland.

In May 2017, evidence of life on land may have been found in 3.48 billion-year-old geyserite which is often found around hot springs and geysers, and other related mineral deposits, uncovered in the Pilbara Craton of Western Australia. This complements the November 2013 publication that microbial mat fossils had been found in 3.48 billion-year-old sandstone in Western Australia.

In November 2017, a study by the University of Edinburgh suggested that life on Earth may have originated from biological particles carried by streams of space dust.

A December 2017 report stated that 3.465-billion-year-old Australian Apex chert rocks once contained microorganisms, the earliest *direct* evidence of life on Earth.

In January 2018, a study found that 4.5 billion-year-old meteorites found on Earth contained liquid water along with prebiotic complex organic substances that may be ingredients for life.

According to biologist Stephen Blair Hedges, "If life arose relatively quickly on Earth ... then it could be common in the universe."

Gallery

Figure 34: *Stromatolites are made by microbes moving upward to avoid being smothered by sediment.*

Figure 35: *Stromatolites left behind by cyanobacteria are one of the oldest fossils of life on Earth.*

Figure 36: *The cyanobacterial-algal mat, salty lake on the White Sea seaside.*

Figure 37: *Wrinkled Kinneyia-type sedimentary structures formed beneath cohesive microbial mats in peritidal zones.*

Figure 38: *Kinneyia-like structure in the Grimsby Formation (Silurian) exposed in Niagara Gorge, NY.*

External links

- Biota[196] (Taxonomicon)
- Life[197] (Systema Naturae 2000)
- Vitae[198] (BioLib)

- Wikispecies – a free directory of life
- Google Images: Earliest known life forms[199]

Conceptual history

Spontaneous generation

Spontaneous generation refers to an obsolete body of thought on the ordinary formation of living organisms without descent from similar organisms. The theory of spontaneous generation held that living creatures could arise from nonliving matter and that such processes were commonplace and regular. For instance, it was hypothesized that certain forms such as fleas could arise from inanimate matter such as dust, or that maggots could arise from dead flesh. A variant idea was that of **equivocal generation**, in which species such as tapeworms arose from unrelated living organisms, now understood to be their hosts. The idea of **univocal generation**, by contrast, refers to effectively exclusive reproduction from genetically related parent(s), generally of the same species.

The doctrine of spontaneous generation was coherently synthesized by Aristotle, who compiled and expanded the work of earlier natural philosophers and the various ancient explanations for the appearance of organisms, and was taken as scientific fact for two millennia. Though challenged in the 17th and 18th centuries by the experiments of Francesco Redi and Lazzaro Spallanzani, spontaneous generation was not disproved until the work of Louis Pasteur and John Tyndall in the mid-19th century.[200]

Rejection of spontaneous generation is no longer controversial among biologists. By the middle of the 19th century, experiments of Louis Pasteur and others refuted the traditional theory of spontaneous generation and supported biogenesis.[201,202]

Figure 39: *Pasteur invented the swan-necked flask to create an environ-*
ment known not to grow microorganisms. After sterilizing a nutrient
broth in these flasks, he removed the swan necks of the controls. Mi-
croorganisms grew only in the controls, refuting spontaneous generation.

Description and terms

Spontaneous generation refers both to the supposed processes by which different types of life might repeatedly emerge from specific sources other than
seeds, eggs, or parents, and also to theoretical principles presented in support
of any such phenomena. Crucial to this doctrine are the ideas that life comes
from non-life and that no causal agent, such as a parent, is needed. The hypothetical processes by which life routinely emerges from nonliving matter
on a time scale of minutes, weeks, or years (e.g. in the supposed seasonal
generation of mice and other animals from the mud of the Nile) are sometimes referred to as *abiogenesis*.[203] Such ideas have no operative principles
in common with the modern hypothesis of abiogenesis, which asserts that life
emerged in the early ages of the planet, over a time span of at least millions
of years, and subsequently diversified, and that there is no evidence of any
subsequent repetition of the event.

The term *equivocal generation*, sometimes known as *heterogenesis* or *xeno-
genesis*, describes the supposed process by which one form of life arises from
a different, unrelated form, such as tapeworms from the bodies of their hosts.

In the years following Louis Pasteur's 1859 experiment, the term "spontaneous generation" fell increasingly out of favor. Experimentalists used a variety of terms for the study of the origin of life from nonliving materials. *Heterogenesis* was applied to the generation of living things from once-living organic matter (such as boiled broths), and Henry Charlton Bastian proposed the term *archebiosis* for life originating from inorganic materials. Disliking the randomness and unpredictability implied by the term "'spontaneous' generation," in 1870 Bastian coined the term *biogenesis* to refer to the formation of life from nonliving matter. Soon thereafter, however, English biologist Thomas Henry Huxley proposed the term *abiogenesis* to refer to this same process and adopted *biogenesis* for the process by which life arises from existing life; it is this latter set of definitions that became dominant.

Antiquity

Presocratic philosophers

Active in the 6th and 5th centuries BCE, early Greek philosophers, called *physiologoi* in antiquity (Greek: φυσιολόγοι; in English, physical or natural philosophers), attempted to give natural explanations of phenomena that had previously been ascribed to the agency of the gods.[204] The *physiologoi* sought the material principle or *arche* (Greek: ἀρχή) of things, emphasizing the rational unity of the external world and rejecting theological or mythological explanations.[205]

Anaximander, who believed that all things arose from the elemental nature of the universe, the *apeiron* (ἄπειρον) or the "unbounded" or "infinite," was likely the first western thinker to propose that life developed spontaneously from nonliving matter. The primal chaos of the *apeiron,* eternally in motion, served as a substratum in which elemental opposites (e.g., wet and dry, hot and cold) generated and shaped the many and varied things in the world. According to Hippolytus of Rome in the third century CE, Anaximander claimed that fish or fish-like creatures were first formed in the "wet" when acted on by the heat of the sun and that these aquatic creatures gave rise to human beings. Censorinus, writing in the 3rd century, reports:

> *Anaximander of Miletus considered that from warmed up water and earth emerged either fish or entirely fishlike animals. Inside these animals, men took form and embryos were held prisoners until puberty; only then, after these animals burst open, could men and women come out, now able to feed themselves.*[206]

Anaximenes, a pupil of Anaximander, thought that air was the element that imparted life and endowed creatures with motion and thought. He proposed

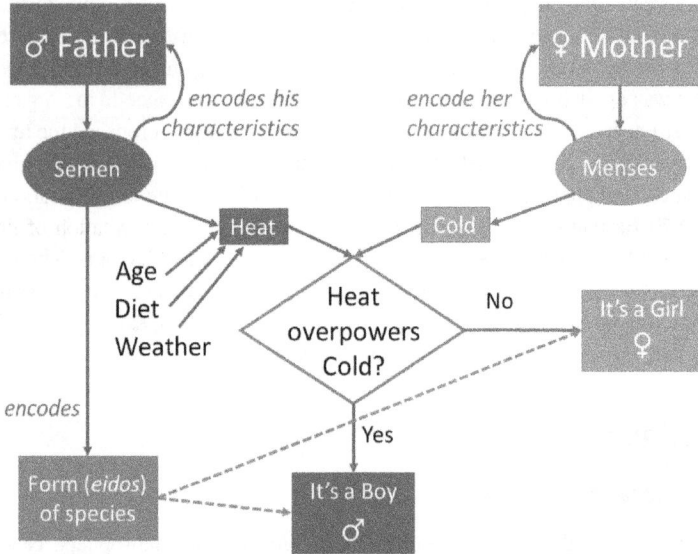

Figure 40: *Aristotle proposed that in sexual reproduction, the child inherits form (eidos) from the father and matter from the mother, as well as heat (pneuma) either from the father or from the environment. In spontaneous generation, the environment could effectively replace the parents' contributions of form, matter, and heat.*

that plants and animals, including human beings, arose from a primordial terrestrial slime, a mixture of earth and water, combined with the sun's heat. Anaxagoras, too, believed that life emerged from a terrestrial slime. However, he held that the seeds of plants existed in the air from the beginning, and those of animals in the aether. Xenophanes traced the origin of man back to the transitional period between the fluid stage of the earth and the formation of land, under the influence of the sun.

In what has occasionally been seen as a prefiguration of a concept of natural selection, Empedocles accepted the spontaneous generation of life but held that different forms, made up of differing combinations of parts, spontaneously arose as though by trial and error: successful combinations formed the species we now see, whereas unsuccessful forms failed to reproduce.

Aristotle

In his biological works, the natural philosopher Aristotle theorized extensively the reproduction of various animals, whether by sexual, parthenogenetic, or spontaneous generation. In accordance with his fundamental theory of hylomorphism, which held that every physical entity was a compound of matter and form, Aristotle's basic theory of sexual reproduction contended that the male's seed imposed form, the set of characteristics passed down to offspring on the "matter" (menstrual blood) supplied by the female. Thus female matter is the material cause of generation—it supplies the matter that will constitute the offspring—while the male semen is the efficient cause, the factor that instigates and delineates the thing's existence. Yet, as proposed in the *History of Animals*, many creatures form not through sexual processes but by spontaneous generation:

> *Now there is one property that animals are found to have in common with plants. For some plants are generated from the seed of plants, whilst other plants are self-generated through the formation of some elemental principle similar to a seed; and of these latter plants some derive their nutriment from the ground, whilst others grow inside other plants ... So with animals, some spring from parent animals according to their kind, whilst others grow spontaneously and not from kindred stock; and of these instances of spontaneous generation some come from putrefying earth or vegetable matter, as is the case with a number of insects, while others are spontaneously generated in the inside of animals out of the secretions of their several organs.*

—*Aristotle, History of Animals, Book V, Part 1*

According to this theory, living things may come forth from nonliving things in a manner roughly analogous to the "enformation of the female matter by the agency of the male seed" seen in sexual reproduction. Nonliving materials, like the seminal fluid present in sexual generation, contain *pneuma* (πνεῦμα, "breath"), or "vital heat". According to Aristotle, pneuma had more "heat" than regular air did, and this heat endowed the substance with certain vital properties:

> *The power of every soul seems to have shared in a different and more divine body than the so called [four] elements ... For every [animal], what makes the seed generative inheres in the seed and is called its "heat." But this is not fire or some such power, but instead the pneuma that is enclosed in the seed and in foamy matter, this being analogous to the element of the stars. This is why fire does not generate any animal ... but the heat of the sun and the heat of animals does, not only the heat that fills the seed, but also any other residue of [the animal's] nature that may exist similarly possesses this vital principle.*

—*Aristotle, Generation of Animals, 736b29ff.*

Figure 41: *Aristotle stated that scallops form spontaneously in sand.*

Aristotle drew an analogy between the "foamy matter" (τὸ ἀφρῶδες) found in nature and the "seed" of an animal, which he viewed as being a kind of foam itself (composed, as it was, from a mixture of water and pneuma). For Aristotle, the generative materials of male and female animals (semen and menstrual blood) were essentially refinements, made by male and female bodies according to their respective proportions of heat, of ingested food, which was, in turn, a byproduct of the elements earth and water. Thus any creature, whether generated sexually from parents or spontaneously through the interaction of vital heat and elemental matter, was dependent on the proportions of pneuma and the various elements which Aristotle believed comprised all things. While Aristotle recognized that many living things emerged from putrefying matter, he pointed out that the putrefaction was not the source of life, but the byproduct of the action of the "sweet" element of water.

Animals and plants come into being in earth and in liquid because there is water in earth, and air in water, and in all air is vital heat so that in a sense all things are full of soul. Therefore living things form quickly whenever this air and vital heat are enclosed in anything. When they are so enclosed, the corporeal liquids being heated, there arises as it were a frothy bubble.

—*Aristotle, Generation of Animals, Book III, Part 11*

With varying degrees of observational confidence, Aristotle theorized the spontaneous generation of a range of creatures from different sorts of inanimate matter. The testaceans (a genus which for Aristotle included bivalves and snails), for instance, were characterized by spontaneous generation from mud, but differed based upon the precise material they grew in—for example, clams and scallops in sand, oysters in slime, and the barnacle and the limpet in the hollows of rocks.

Latin and early Christian sources

Vitruvius, a Roman architect and writer of the 1st century BCE, advised that libraries be placed facing eastwards to benefit from morning light, but not towards the south or the west as those winds generate bookworms.

Aristotle claimed that eels were lacking in sex and lacking milt, spawn and the passages for either. Rather, he asserted eels emerged from earthworms. Later authors dissented. Pliny the Elder did not argue against the anatomic limits of eels, but stated that eels reproduce by budding, scraping themselves against rocks, liberating particles that become eels. Athenaeus described eels as entwining and discharging a fluid which would settle on mud and generate life. On the other hand, Athenaeus also dissented towards spontaneous generation, claiming that a variety of anchovy did not generate from roe, as Aristotle stated, but rather, from sea foam.

As the dominant view of philosophers and thinkers continued to be in favour of spontaneous generation, some Christian theologians accepted the view. Augustine of Hippo discussed spontaneous generation in *The City of God* and *The Literal Meaning of Genesis*, citing Biblical passages such as "Let the waters bring forth abundantly the moving creature that hath life" (Genesis 1:20) as decrees that would enable ongoing creation.

Middle Ages

From the fall of the Roman Empire in 5th century to the East-West Schism in 1054, the influence of Greek science declined, although spontaneous generation generally went unchallenged. New descriptions were made. Of the numerous beliefs, some had doctrinal implications outside of the Book of Genesis. For example, the idea that a variety of bird known as the *barnacle goose* emerged from a crustacean known as the *goose barnacle*, had implications on the practice of fasting during Lent. In 1188, Gerald of Wales, after having traveled in Ireland, argued that the "unnatural" generation of barnacle geese was evidence for the virgin birth. Where the practice of fasting during Lent allowed fish, but prohibited fowl, the idea that the goose was in fact a fish suggested that its consumption be permitted during Lent. The practice was eventually prohibited by decree of Pope Innocent III in 1215.

Aristotle, in Arabic translation, was reintroduced to Western Europe. During the 13th century, Aristotle reached his greatest acceptance. With the availability of Latin translations Saint Albertus Magnus and his student, Saint Thomas

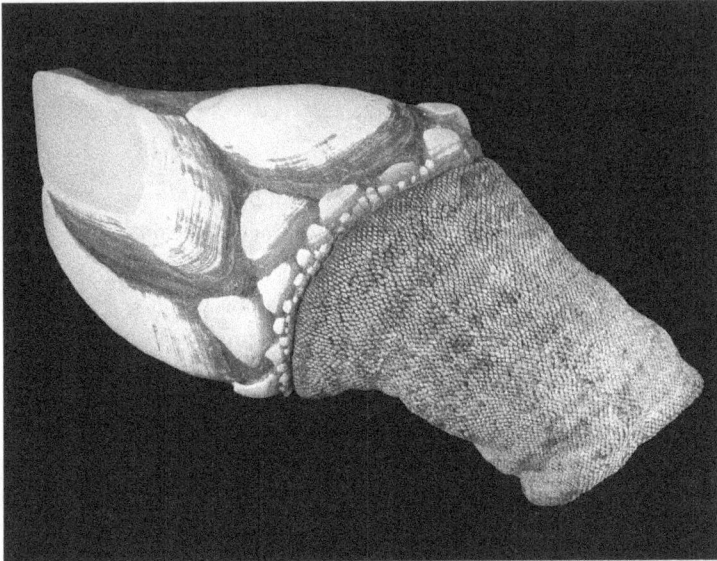

Figure 42: *The goose barnacle:*
Pollicipes cornucopia

Figure 43: *The barnacle goose:*
Branta leucopsis

Aquinas, raised Aristotelianism to its greatest prominence. Albert wrote a paraphrase of Aristotle, *De causis et processu universitatis*, in which he removed some and incorporated other commentaries by Arabic scholars. The influential writings of Aquinas, on both the physical and metaphysical, are predominantly Aristotelian, but show numerous other influences.

Spontaneous generation is discussed as a fact in literature well into the Renaissance. Where, in passing, Shakespeare discusses snakes and crocodiles forming from the mud of the Nile (Ant 2.7 F1[207]), Izaak Walton again raises the question of the origin of eels "as rats and mice, and many other living creatures, are bred in Egypt, by the sun's heat when it shines upon the overflowing of the river...". While the ancient question of the origin of eels remained unanswered and the additional idea that eels reproduced from corruption of age was mentioned, the spontaneous generation of rats and mice engendered no debate.

The Dutch biologist and microscopist Jan Swammerdam (1637 - 1680) rejected the concept that one animal could arise from another or from putrification by chance because it was impious and like others found the concept of spontaneous generation irreligious, and he associated it with atheism and Godless opinion.

Modern tests

Jan Baptist van Helmont (1580–1644) used experimental techniques, such as growing a willow for five years and showing it increased mass while the soil showed a trivial decrease in comparison. As the process of photosynthesis was not understood, he attributed the increase of mass to the absorption of water. His notes also describe a recipe for mice (a piece of soiled cloth plus wheat for 21 days) and scorpions (basil, placed between two bricks and left in sunlight). His notes suggest he may even have done these things.

Where Aristotle held that the embryo was formed by a coagulation in the uterus, William Harvey (1578 - 1657) by way of dissection of deer, showed that there was no visible embryo during the first month. Although his work predated the microscope, this led him to suggest that life came from invisible eggs. In the frontispiece of his book *Exercitationes de Generatione Animalium* (*Essays on the Generation of Animals*), he made an expression of biogenesis: "omnia ex ovo" (everything from eggs).

The ancient beliefs were subjected to testing. In 1668, Francesco Redi challenged the idea that maggots arose spontaneously from rotting meat. In the first major experiment to challenge spontaneous generation, he placed meat in a variety of sealed, open, and partially covered containers. Realizing that the sealed containers were deprived of air, he used "fine Naples veil", and observed no worm on the meat, but they appeared on the cloth. Redi used his

Figure 44: *A modern rendering of Francesco Redi's 1668 experiment on abiogenesis. Flies form on the meat in the open jar (left) but not in the closed jar (right).*

experiments to support the preexistence theory put forth by the Church at that time, which maintained that living things originated from parents. In scientific circles Redi's work very soon had great influence, as evidenced in a letter from John Ray in 1671 to members of the Royal Society of London:

Whether there be any spontaneous or anomalous generation of animals, as has been the constant opinion of naturalists heretofore, I think there is good reason to question. It seems to me at present most probable, that there is no such thing; but that even all insects are the natural issue of parents of the same species with themselves. F. Redi has gone a good way in proving this, having cleared the point concerning generation ex materia putrida. But still there remain two great difficulties. The first is, to give an account of the production of insects bred in the by-fruits and excrescencies of vegetables, which the said Redi doubts not to ascribe to the vegetative soul of the plant that yields those excrescencies. But for this I refer you to Mr. Lister. The second, to render an account of insects bred in the bodies of other animals. I hope shortly to be able to give you an account of the generation of some of those insects which have been thought to be spontaneous, and which seem as unlikely as any to be after the ordinary and usual way.[208]

Pier Antonio Micheli, around 1729, observed that when fungal spores were placed on slices of melon the same type of fungi were produced that the spores came from, and from this observation he noted that fungi did not arise from spontaneous generation.

In 1745, John Needham performed a series of experiments on boiled broths. Believing that boiling would kill all living things, he showed that when sealed right after boiling, the broths would cloud, allowing the belief in spontaneous generation to persist. His studies were rigorously scrutinized by his peers and many of them agreed.

Lazzaro Spallanzani modified the Needham experiment in 1768, attempting to exclude the possibility of introducing a contaminating factor between boiling and sealing. His technique involved boiling the broth in a sealed container with the air partially evacuated to prevent explosions. Although he did not see growth, the exclusion of air left the question of whether air was an essential factor in spontaneous generation. However, by that time there was already widespread scepticism among major scientists, to the principle of spontaneous generation. Observation was increasingly demonstrating that whenever there was sufficiently careful investigation of mechanisms of biological reproduction, it was plain that processes involved basing of new structures on existing complex structures, rather from chaotic muds or dead materials. Joseph Priestley, after he had fled to America and not long before his death, wrote a letter that was read to the American Philosophical Society in 1803. It said in part:

There is nothing in modern philosophy that appears to me so extraordinary, as the revival of what has long been considered as the exploded doctrine of equivocal, or, as Dr. Darwin calls it, spontaneous generation; by which is meant the production of organized bodies from substances that have no organization, as plants and animals from no pre-existing germs of the same kinds, plants without seeds, and animals without sexual intercourse.

The germ of an organized body, the seed of a plant, or the embrio of an animal, in its first discoverable state, is now found to be the future plant or animal in miniature, containing every thing essential to it when full grown, only requiring to have the several organs enlarged, and the interstices filled with extraneous nutritive matter. When the external form undergoes the greatest change, as from an aquatic insect to a flying gnat, a caterpillar to a crysalis, a crysalis to a butterfly, or a tadpole to a frog, there is nothing new in the organization; all the parts of the gnat, the butterfly, and the frog, having really existed, though not appearing to the common observer in the forms in which they are first seen. In like manner, every thing essential to the oak is found in the acorn.[209]

In 1837, Charles Cagniard de la Tour, a physicist, and Theodor Schwann, one of the founders of cell theory, published their independent discovery of yeast in alcoholic fermentation. They used the microscope to examine foam left over from the process of brewing beer. Where Leeuwenhoek described "small spheroid globules", they observed yeast cells undergo cell division. Fermentation would not occur when sterile air or pure oxygen was introduced if yeast were not present. This suggested that airborne microorganisms, not spontaneous generation, was responsible.

However, although the idea of spontaneous generation had been in decline for nearly a century, its supporters did not abandon it all at once. As James Rennie wrote:

> ...inability to trace the origin of minute plants and insects led to the doctrine of what is called spontaneous or equivocal generation, of which the fancies above-mentioned are some 'of the prominent branches. The experiments of Redi on the hatching of insects from eggs, which were published at Florence in 1668, first brought discredit upon this doctrine, though it had always a few eminent disciples. At present it is maintained by a considerable number -of distinguished naturalists, such as Blumenbach, Cuvier, Bory de St. Vincent, R. Brown, &c. 'The notion or spontaneous generation, "says Bory, "is at first revolting to a rational mind, but it is, notwithstanding, demonstrable by the microscope. The fact is averred : Willer has seen it, I have seen it, and twenty other observers have seen it: the pandorinia exhibit it every instant. 'These pandorinia he elsewhere describes as probably nothing more than "animated scions of Zoocarpae". It would be unprofitable to go into any lengthened discussion upon this mysterious subject; and we have great doubts whether the ocular demonstration by the microscope would succeed except in the hands of a disciple of the school. Even with naturalists, whose business it is to deal with facts, the reason is often wonderfully influenced by the imagination...[210]

Pasteur and Tyndall

Louis Pasteur's 1859 experiment is widely seen as having settled the question of spontaneous generation. He boiled a meat broth in a flask that had a long neck that curved downward, like that of a goose or swan. The idea was that the bend in the neck prevented falling particles from reaching the broth, while still allowing the free flow of air. The flask remained free of growth for an extended period. When the flask was turned so that particles could fall down the bends, the broth quickly became clouded. However, minority objections were persistent and not always unreasonable, given that the experimental difficulties were far more challenging than the popular accounts suggest. The investigations of John Tyndall, a correspondent of Pasteur and a great admirer of Pasteur's work, were decisive in disproving spontaneous generation and dealing with lingering issues. Still, even Tyndall encountered difficulties in dealing with the effects of microbial spores, which were not well understood in his day. Like Pasteur, he boiled his cultures to sterilize them, and some types of bacterial spores can survive boiling. The autoclave, which eventually came into universal application in medical practice and microbiology to sterilise equipment, was not an instrument that had come into use at the time of Tyndall's experiments, let alone those of Pasteur.

In 1862, the French Academy of Sciences paid a special attention to the issue and established a prize "to him who by well-conducted experiments throws new light on the question of the so-called spontaneous generation" and appointed a commission to judge the winner.

Biogenesis

Life timeline

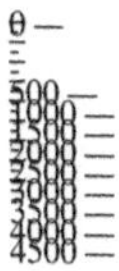

θ —

500
1000
1500
2000
2500
3000
3500
4000
4500

Axis scale: million years

Also see: *Human timeline* and *Nature timeline*

Biogenesis is the production of new living organisms or organelles. Conceptually, biogenesis is primarily attributed to Louis Pasteur and encompasses the belief that complex living things come only from other living things, by means of reproduction. That is, life does not spontaneously arise from non-living material, which was the position held by spontaneous generation.[211,212] This is summarized in the phrase *Omne vivum ex vivo*, Latin for *"all life [is] from life."* A related statement is *Omnis cellula e cellula*, "all cells [are] from cells;" this conclusion is one of the central statements of cell theory.

Biogenesis and abiogenesis

The term biogenesis was coined by Henry Charlton Bastian to mean the generation of a life form from nonliving materials, however, Thomas Henry Huxley chose the term abiogenesis and redefined biogenesis for life arising from preexisting life. The generation of life from non-living material is called abiogenesis, and occurred through stepwise chemical and molecular evolution over millions of years.

The term *biogenesis* may also refer to biochemical processes of production in living organisms (see biosynthesis).

Spontaneous generation and its disproof

The Ancient Greeks believed that living things could spontaneously come into being from nonliving matter, and that the goddess Gaia could make life arise spontaneously from stones – a process known as *Generatio spontanea*. Aristotle disagreed, but he still believed that creatures could arise from dissimilar organisms or from soil. Variations of this concept of spontaneous generation still existed as late as the 17th century, but towards the end of the 17th century, a series of observations and arguments began that eventually discredited such ideas. This advance in scientific understanding was met with much opposition, with personal beliefs and individual prejudices often obscuring the facts.

William Harvey (1578–1657) was an early proponent of all life beginning from an egg, *omne vivum ex ovo*. Francesco Redi, an Italian physician, proved as early as 1668 that higher forms of life did not originate spontaneously by demonstrating that maggots come from eggs of flies. But proponents of spontaneous generation claimed that this did not apply to microbes and continued to hold that these could arise spontaneously. Attempts to disprove the spontaneous generation of life from non-life continued in the early 19th century with observations and experiments by Franz Schulze and Theodor Schwann. In 1745, John Needham added chicken broth to a flask and boiled it. He then let it cool and waited. Microbes grew, and he proposed it as an example of spontaneous generation. In 1768, Lazzaro Spallanzani repeated Needham's experiment but removed all the air from the flask. No growth occurred.[213] In 1854, Heinrich G. F. Schröder (1810–1885) and Theodor von Dusch, and in 1859, Schröder alone, repeated the Helmholtz filtration experiment and showed that living particles can be removed from air by filtering it through cotton-wool.

In 1864, Louis Pasteur finally announced the results of his scientific experiments. In a series of experiments similar to those performed earlier by Needham and Spallanzani, Pasteur demonstrated that life does not arise in areas that have not been contaminated by existing life. Pasteur's empirical results were summarized in the phrase *Omne vivum ex vivo*, Latin for "all life [is] from life".[214,215]

After obtaining his results, Pasteur stated: *"La génération spontanée est une chimère"* ("Spontaneous generation is a dream").

Primordial soup

Primordial soup, or **prebiotic soup**, is a hypothetical condition of the Earth's atmosphere before the emergence of life. It is a chemical environment in which the first biological molecules (organic compounds) were formed under natural forces. According to the theory, simple organic compounds were created from non-living inorganic molecules (abiogenesis) through physical and chemical reactions on the Earth's surface. The so formed organic molecules accumulate into a rich organic ocean, or a "soup". In this soup, simple organic molecules reacted with each other (polymerise) to form more complex molecules, including nucleic acids and proteins, which are the central structural and functional components of all organisms. These molecules then aggregate to become the first forms of life.

The British naturalist Charles Darwin had vaguely imagined the primordial soup as a "warm little pond" in 1871. A coherent scientific argument was introduced by a Soviet biochemist Alexander Oparin in 1924. According to Oparin, in the primitive Earth's surface, carbon, hydrogen, water vapour, and ammonia reacted to form the first organic compounds. Unbeknownst to Oparin, whose writing was circulated only in Russian, an English scientist John Burdon Sanderson Haldane independently arrived at similar conclusion in 1929. It was Haldane who gave the name "soup" to the theory.

The theory is variously known as "primordial soup theory", "prebiotic soup theory", and "Oparin-Haldane hypothesis". Biochemist Robert Shapiro has summarized the theory in its "mature form" as follows:

1. Early Earth had a chemically reducing atmosphere.
2. This atmosphere, exposed to energy in various forms, produced simple organic compounds ("monomers").
3. These compounds accumulated in a "soup", which may have been concentrated at various locations (shorelines, oceanic vents etc.).
4. By further transformation, more complex organic polymers – and ultimately life – developed in the soup.

Historical background

The notion that living beings originated from inanimate materials originated among the Ancient Greeks—the theory known as spontaneous generation. Aristotle in the 4th century BCE gave a proper explanation, writing:

> So with animals, some spring from parent animals according to their kind, whilst others grow spontaneously and not from kindred stock; and of these instances of spontaneous generation some come from putrefying earth or

vegetable matter, as is the case with a number of insects, while others are spontaneously generated in the inside of animals out of the secretions of their several organs.

—Aristotle, On the History of Animals, Book V, Part 1

He also states that it is not only that animals originate from other similar animals, but also that living things do arise and always have arisen from lifeless matter. His theory remained the dominant idea on origin of life from the ancient philosophers to the Renaissance thinkers in various forms. But with the birth of modern science, experimental refutations emerged. An Italian physician Francesco Redi demonstrated in 1668 that maggots developed from rotten meat only in a jar where flies could enter, but not in closed-lid jar. He concluded that: *omne vivum ex vivo* (All life comes from life).

The experiment of a French chemist Louis Pasteur in 1859 is regarded as the death blow to spontaneous generation. He experimentally showed that organisms (microbes) can not grow in a sterilised water, unless it is exposed to air. The experiment won him the Alhumbert Prize in 1862 from the French Academy of Sciences, and he concluded: Never will the doctrine of spontaneous generation recover from the mortal blow of this simple experiment.

But evolutionary biologists believed that a kind of spontaneous generation, but different from the simple Aristotelian doctrine, must have worked for the emergence of life. A French biologist Jean-Baptiste de Lamarck had speculated that the first life form started from non-living materials. "Nature, by means of heat, light, electricity and moisture, he wrote in 1809 in *Philosophie Zoologique (The Philosophy of Zoology)*, "forms direct or spontaneous generation at that extremity of each kingdom of living bodies, where the simplest of these bodies are found." When an English naturalist Charles Darwin introduced his theory of natural selection in his book *On the Origin of Species* in 1859, and even in his subsequent books, his supporters, such as a German zoologist Ernst Haeckel, criticised him for not using his theory to explain the origin of life. Haeckel wrote in 1862: "The chief defect of the Darwinian theory is that it throws no light on the origin of the primitive organism—probably a simple cell—from which all the others have descended. When Darwin assumes a special creative act for this first species, he is not consistent, and, I think, not quite sincere."

But Darwin wrote (in a personal letter to Joseph Dalton Hooker) in 1871, expressing his idea on the origin of life as:

It is often said that all the conditions for the first production of a living organism are now present, which could ever have been present.— But if (& oh what a big if) we could conceive in some warm little pond with all sorts of ammonia & phosphoric salts,—light, heat, electricity &c present,

*that a protein compound was chemically formed, ready to undergo still
more complex changes, at the present day such matter w^d be instantly
devoured, or absorbed, which would not have been the case before living
creatures were formed.*

Oparin's theory

Alexander Oparin first postulated his theory in Russian in 1924 in a small
pamphlet titled *Proiskhozhdenie Zhizny* (*The Origin of Life*). According to
Oparin, the primitive Earth's surface had a thick red-hot liquid, composed of
heavy elements such as carbon (in the form of iron carbide). This nucleus was
surrounded by lightest elements, i.e. gases, such as hydrogen. In the presence of water vapour, carbides reacted with hydrogen to form hydrocarbons.
Such hydrocarbons were the first organic molecules. These further combined
with oxygen and ammonia to produce hydroxy- and amino-derivatives, such
as carbohydrates and proteins. These molecules accumulated on the ocean's
surface, becoming gel-like substances and growing in size. They gave rise to
primitive organisms (cells). In his original theory, Oparin considered oxygen
as one of the primordial gases; thus the primordial atmosphere was an oxi-
dising one. However, when he elaborated his theory in 1936 (in a book by
the same title, and translated into English in 1938), he modified the chemical
composition of the primordial environment as strictly reducing, consisting of
methane, ammonia, free hydrogen and water vapour—excluding oxygen.

Haldane's theory

J.B.S. Haldane independently postulated his primordial soup theory in 1929 in
an eight-page article "The origin of life" in *The Rationalist Annual*. According
to Haldane the primitive Earth's atmosphere was essentially reducing, with little or no oxygen. Ultraviolet ray from the Sun-induced reaction on a mixture
of water, carbon dioxide, and ammonia. Organic substances such as sugars and
protein components (amino acids) were synthesised. These molecules "accumulated till the primitive oceans reached the consistency of hot dilute soup."
The first reproducing things were created from this soup.

As to the priority over the theory, Haldane accepted that Oparin came first,
saying, "I have very little doubt that Professor Oparin has the priority over
me."

Monomer formation

One of the most important pieces of experimental support for the "soup" theory came in 1953. A graduate student, Stanley Miller, and his professor, Harold Urey, performed an experiment that demonstrated how organic molecules could have spontaneously formed from inorganic precursors, under conditions like those posited by the Oparin-Haldane Hypothesis. The now-famous "Miller–Urey experiment" used a highly reduced mixture of gases—methane, ammonia and hydrogen—to form basic organic monomers, such as amino acids. This provided direct experimental support for the second point of the "soup" theory, and it is around the remaining two points of the theory that much of the debate now centers.

Apart from the Miller–Urey experiment, the next most important step in research on prebiotic organic synthesis was the demonstration by Joan Oró that the nucleic acid purine base, adenine, was formed by heating aqueous ammonium cyanide solutions. In support of abiogenesis in eutectic ice, more recent work demonstrated the formation of s-triazines (alternative nucleobases), pyrimidines (including cytosine and uracil), and adenine from urea solutions subjected to freeze-thaw cycles under a reductive atmosphere (with spark discharges as an energy source).

Further transformation

The spontaneous formation of complex polymers from abiotically generated monomers under the conditions posited by the "soup" theory is not at all a straightforward process. Besides the necessary basic organic monomers, compounds that would have prohibited the formation of polymers were formed in high concentration during the Miller–Urey and Oró experiments.Wikipedia:Citation needed The Miller experiment, for example, produces many substances that would undergo cross-reactions with the amino acids or terminate the peptide chain.Wikipedia:Citation needed

Proteinoid

Proteinoids, or **thermal proteins**, are protein-like, often cross-linked molecules formed abiotically from amino acids. Sidney W. Fox initially proposed that they may have been precursors to the first living cells (protocells). The term was also in the 1960s to describe peptides that are shorter than twenty amino acids found in hydrolysed protein, but this term is no longer commonly used.

History

In trying to uncover the intermediate stages of abiogenesis, scientist Sidney W. Fox in the 1950s and 1960s, studied the spontaneous formation of peptide structures under conditions that might plausibly have existed early in Earth's history. He demonstrated that amino acids could spontaneously form small chains called peptides. In one of his experiments, he allowed amino acids to dry out as if puddled in a warm, dry spot in prebiotic conditions. He found that, as they dried, the amino acids formed long, often cross-linked, thread-like microscopic polypeptide globules, he named "proteinoid microspheres".[216]

Polymerization

The abiotic polymerization of amino acids into proteins through the formation of peptide bonds was thought to occur only at temperatures over 140 °C. However, the biochemist Sidney Walter Fox and his co-workers discovered that phosphoric acid acted as a catalyst for this reaction.Wikipedia:Citation needed They were able to form protein-like chains from a mixture of 18 common amino acids at 70 °C in the presence of phosphoric acid, and dubbed these protein-like chains proteinoids. Fox later found naturally occurring proteinoids similar to those he had created in his laboratory in lava and cinders from Hawaiian volcanic vents and determined that the amino acids present polymerized due to the heat of escaping gases and lava.Wikipedia:Citation needed Other catalysts have since been found; one of them, amidinium carbodiimide, is formed in primitive Earth experiments and is effective in dilute aqueous solutions.

When present in certain concentrations in aqueous solutions, proteinoids form small microspheres. This is because some of the amino acids incorporated into proteinoid chains are more hydrophobic than others, and so proteinoids cluster together like droplets of oil in water. These structures exhibit a few characteristics of living cells:

1. An outer wall.

2. Osmotic swelling and shrinking.Wikipedia:Citation needed
3. Budding.Wikipedia:Citation needed
4. Binary fission (dividing into two daughter microspheres).Wikipedia:Citation needed
5. Streaming movement of internal particles.Wikipedia:Citation needed

Fox thought that the microspheres may have provided a cell compartment within which organic molecules could have become concentrated and protected from the outside environment during the process of chemical evolution.

Proteinoid microspheres are today being considered for use in pharmaceuticals, providing microscopic biodegradable capsules in which to package and deliver oral drugs.[217]

In another experiment using a similar method to set suitable conditions for life to form, Fox collected volcanic material from a cinder cone in Hawaii. He discovered that the temperature was over 100 °C (212 °F) just 4 inches (100 mm) beneath the surface of the cinder cone, and suggested that this might have been the environment in which life was created—molecules could have formed and then been washed through the loose volcanic ash and into the sea.Wikipedia:Citation needed He placed lumps of lava over amino acids derived from methane, ammonia and water, sterilized all materials, and baked the lava over the amino acids for a few hours in a glass oven. A brown, sticky substance formed over the surface and when the lava was drenched in sterilized water a thick, brown liquid leached out. It turned out that the amino acids had combined to form proteinoids, and the proteinoids had combined to form small spheres. Fox called these "microspheres". His protobionts were not cells, although they formed clumps and chains reminiscent of bacteria. Based upon such experiments, Colin Pittendrigh stated in December 1967 that "laboratories will be creating a living cell within ten years," a remark that reflected the typical contemporary levels of ignorance of the complexity of cell structures.

Legacy

Fox has likened the amino acid globules to cells, and proposed it bridged the macromolecule to cell transition. However, his hypothesis was later dismissed as proteinoids are not proteins, they feature mostly non-peptide bonds and amino acid cross-linkages not present in living organisms. Furthermore, they have no compartmentalization and there is no information content in the molecules.

Although their role as an evolutionary precursor has been superseded, the hypothesis was a catalyst to further investigate other mechanisms that could have brought about abiogenesis, such as the RNA world, PAH world, Iron–sulfur world, and protocell hypotheses.Wikipedia:Citation needed

Further reading

• Fox, Sidney W.; Kaoru Harada (14 November 1958). "Thermal Copoly-
  merization of Amino Acids to a Product Resembling Protein". *Science*.
  New Series. **128** (3333): 1214. Bibcode: 1958Sci...128.1214F[218]. doi:
  10.1126/science.128.3333.1214[219]. JSTOR 1756313[220].

Autocatalysis

Autocatalysis

A single chemical reaction is said to be **autocatalytic** if one of the reaction products is also a catalyst for the same or a coupled reaction.[221] Such a reaction is called an **autocatalytic reaction**.

A *set* of chemical reactions can be said to be "collectively autocatalytic" if a number of those reactions produce, as reaction products, catalysts for enough of the other reactions that the entire set of chemical reactions is self-sustaining given an input of energy and food molecules (see autocatalytic set).

Chemical reactions

A chemical reaction of two reactants and two products can be written as

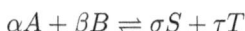

$$\alpha A + \beta B \rightleftharpoons \sigma S + \tau T$$

where the Greek letters are stoichiometric coefficients and the capital Latin letters represent chemical species. The chemical reaction proceeds in both the forward and reverse direction. This equation is easily generalized to any number of reactants, products, and reactions.

Chemical equilibrium

In chemical equilibrium the forward and reverse reaction rates are such that each chemical species is being created at the same rate it is being destroyed. In other words, the rate of the forward reaction is equal to the rate of the reverse reaction.

$$k_+[A]^\alpha[B]^\beta = k_-[S]^\sigma[T]^\tau$$

Here, the brackets indicate the concentration of the chemical species, in moles per liter, and k_+ and k_- are rate constants.

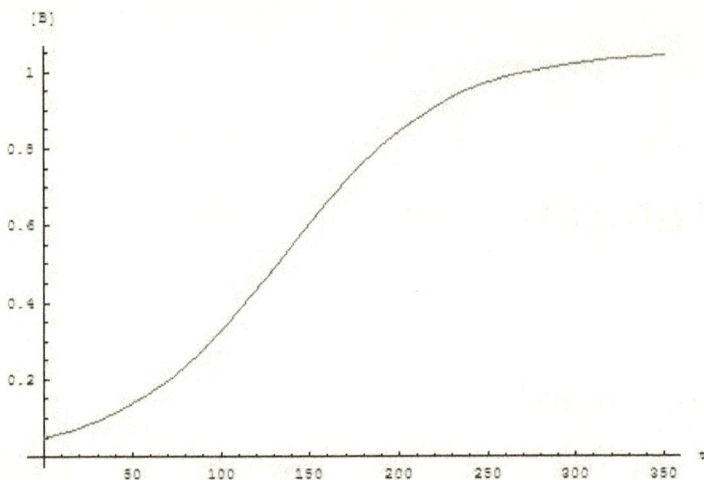

Figure 45: *Sigmoid variation of product concentration in autocatalytic reactions*

Far from equilibrium

Far from equilibrium, the forward and reverse reaction rates no longer balance and the concentration of reactants and products is no longer constant. For every forward reaction α molecules of A are destroyed. For every reverse reaction α molecules of A are created. In the case of an elementary reaction step the reaction order in each direction equals the molecularity, so that the rate of change in number of moles of A is then

$$\frac{d}{dt}[A] = -\alpha k_+[A]^\alpha[B]^\beta + \alpha k_-[S]^\sigma[T]^\tau$$

$$\frac{d}{dt}[B] = -\beta k_+[A]^\alpha[B]^\beta + \beta k_-[S]^\sigma[T]^\tau$$

$$\frac{d}{dt}[S] = \sigma k_+[A]^\alpha[B]^\beta - \sigma k_-[S]^\sigma[T]^\tau$$

$$\frac{d}{dt}[T] = \tau k_+[A]^\alpha[B]^\beta - \tau k_-[S]^\sigma[T]^\tau$$

This system of equations has a single stable fixed point when the forward rates and the reverse rates are equal. This means that the system evolves to the equilibrium state, and this is the only state to which it evolves.

Autocatalytic reactions

Autocatalytic reactions are those in which at least one of the products is a reactant. Perhaps the simplest autocatalytic reaction can be written

$$A + B \rightleftharpoons 2B$$

with the rate equations (for an elementary reaction)

$$\frac{d}{dt}[A] = -k_+[A][B] + k_-[B]^2$$

$$\frac{d}{dt}[B] = +k_+[A][B] - k_-[B]^2 .$$

This reaction is one in which a molecule of species A interacts with a molecule of species B. The A molecule is converted into a B molecule. The final product consists of the original B molecule plus the B molecule created in the reaction.

The key feature of these rate equations is that they are nonlinear; the second term on the right varies as the square of the concentration of B. This feature can lead to multiple fixed points of the system, much like a quadratic equation can have multiple roots. Multiple fixed points allow for multiple states of the system. A system existing in multiple macroscopic states is more orderly (has lower entropy) than a system in a single state.

The concentrations of A and B vary in time according to[222]

$$[A] = \frac{[A]_0 + [B]_0}{1 + \frac{[B]_0}{[A]_0} e^{([A]_0 + [B]_0)kt}}$$

and

$$[B] = \frac{[A]_0 + [B]_0}{1 + \frac{[A]_0}{[B]_0} e^{-([A]_0 + [B]_0)kt}} .$$

The graph for these equations is a sigmoid curve, which is typical for auto-catalytic reactions: these chemical reactions proceed slowly at the start (the induction period) because there is little catalyst present, the rate of reaction increases progressively as the reaction proceeds as the amount of catalyst increases and then it again slows down as the reactant concentration decreases. If the concentration of a reactant or product in an experiment follows a sigmoid curve, the reaction may be autocatalytic.

These kinetic equations apply for example to the acid-catalyzed hydrolysis of some esters to carboxylic acids and alcohols. There must be at least some acid present initially to start the catalyzed mechanism; if not the reaction must start by an alternate uncatalyzed path which is usually slower. The above equations for the catalyzed mechanism would imply that the concentration of acid product remains zero forever.

Creation of order

Background

The Second Law of Thermodynamics states that the disorder (entropy) of a physical or chemical system and its surroundings (a closed system) must increase with time. Systems left to themselves become increasingly random, and orderly energy of a system like uniform motion degrades eventually to the random motion of particles in a heat bath.

There are, however, many instances in which physical systems spontaneously become emergent or orderly. For example, despite the destruction they cause, hurricanes have a very orderly vortex motion when compared to the random motion of the air molecules in a closed room. Even more spectacular is the order created by chemical systems; the most dramatic being the order associated with life.

This is consistent with the Second Law, which requires that the total disorder of a system *and its surroundings* must increase with time. Order can be created in a system by an even greater decrease in order of the system's surroundings. In the hurricane example, hurricanes are formed from unequal heating within the atmosphere. The Earth's atmosphere is then far from thermal equilibrium. The order of the Earth's atmosphere increases, but at the expense of the order of the sun. The sun is becoming more disorderly as it ages and throws off light and material to the rest of the universe. The total disorder of the sun and the earth increases despite the fact that orderly hurricanes are generated on earth.

A similar example exists for living chemical systems. The sun provides energy to green plants. The green plants are food for other living chemical systems. The energy absorbed by plants and converted into chemical energy generates a system on earth that is orderly and far from chemical equilibrium. Here, the difference from chemical equilibrium is determined by an excess of reactants over the equilibrium amount. Once again, order on earth is generated at the expense of entropy increase of the sun. The total entropy of the earth and the rest of the universe increases, consistent with the Second Law.

Some autocatalytic reactions also generate order in a system at the expense of its surroundings. For example, (clock reactions) have intermediates whose concentrations oscillate in time, corresponding to temporal order. Other reactions generate spatial separation of chemical species corresponding to spatial order. More complex reactions are involved in metabolic pathways and metabolic networks in biological systems.

The transition to order as the distance from equilibrium increases is not usually continuous. Order typically appears abruptly. The threshold between the disorder of chemical equilibrium and order is known as a phase transition. The

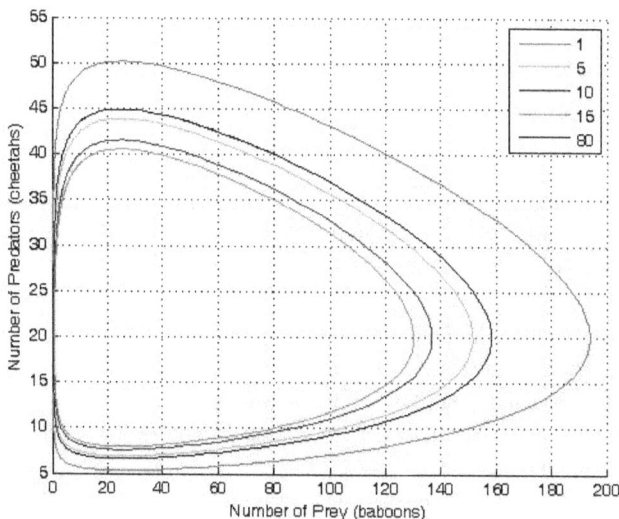

Figure 46: *The Lotka-Volterra equation is isomorphic with the predator prey model and the two reaction autocatalytic model. In this example baboons and cheetahs are equivalent to two different chemical species X and Y in autocatalytic reactions.*

conditions for a phase transition can be determined with the mathematical machinery of non-equilibrium thermodynamics.

Temporal order

A chemical reaction cannot oscillate about a position of final equilibrium because the second law of thermodynamics requires that a thermodynamic system approach equilibrium and not recede from it. For a closed system at constant temperature and pressure, the Gibbs free energy must decrease continuously and not oscillate. However it is possible that the concentrations of some reaction intermediates oscillate, and also that the *rate* of formation of products oscillates.[223]

Idealized example: Lotka-Volterra equation

Consider a coupled set of two autocatalytic reactions in which the concentration of one of the reactants A is much larger than its equilibrium value. In this case the forward reaction rate is so much larger than the reverse rates that we can neglect the reverse rates.

$$A + X \rightarrow 2X$$

$$X + Y \rightarrow 2Y$$
$$Y \rightarrow E$$

with the rate equations

$$\frac{d}{dt}[X] = k_1[A][X] - k_2[X][Y]$$
$$\frac{d}{dt}[Y] = k_2[X][Y] - k_3[Y].$$

Here, we have neglected the depletion of the reactant A, since its concentration is so large. The rate constants for the three reactions are k_1, k_2, and k_3, respectively.

This system of rate equations is known as the Lotka-Volterra equation and is most closely associated with population dynamics in predator-prey relationships. This system of equations can yield oscillating concentrations of the reaction intermediates X and Y. The amplitude of the oscillations depends on the concentration of A (which decreases without oscillation). Such oscillations are a form of emergent temporal order that is not present in equilibrium.

Another idealized example: Brusselator

Another example of a system that demonstrates temporal order is the Brusselator (see Prigogine reference). It is characterized by the reactions

$$A \rightarrow X$$
$$2X + Y \rightarrow 3X$$
$$B + X \rightarrow Y + D$$
$$X \rightarrow E$$

with the rate equations

$$\frac{d}{dt}[X] = [A] + [X]^2[Y] - [B][X] - [X]$$
$$\frac{d}{dt}[Y] = [B][X] - [X]^2[Y]$$

where, for convenience, the rate constants have been set to 1.

The Brusselator has a fixed point at

$$[X] = A$$
$$[Y] = \frac{B}{A}.$$

The fixed point becomes unstable when

$$B > 1 + A^2$$

leading to an oscillation of the system. Unlike the Lotka-Volterra equation, the oscillations of the Brusselator do not depend on the amount of reactant present initially. Instead, after sufficient time, the oscillations approach a limit cycle.[224]

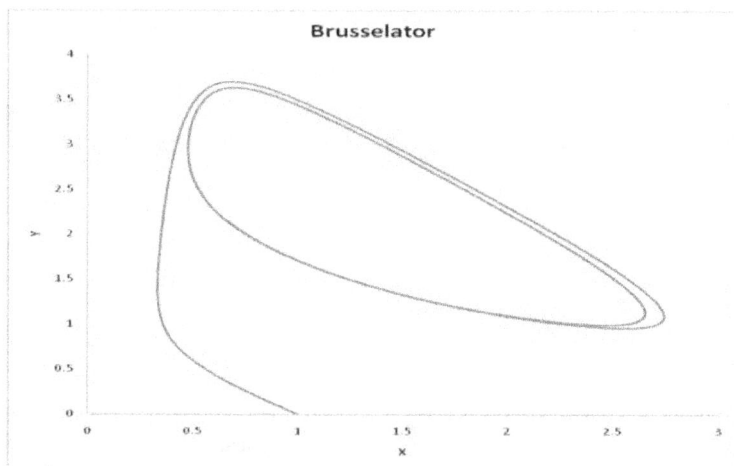

Figure 47: *The Brusselator in the unstable regime. A=1.*
B=2.5. X(0)=1. Y(0)=0. The system approaches a limit cycle.
For B<1+A the system is stable and approaches a fixed point.

Spatial order

An idealized example of spatial spontaneous symmetry breaking is the case in which we have two boxes of material separated by a permeable membrane so that material can diffuse between the two boxes. It is assumed that identical Brusselators are in each box with nearly identical initial conditions. (see Prigogine reference)

$$\frac{d}{dt}[X_1] = [A] + [X_1]^2[Y_1] - [B][X_1] - [X_1] + D_x(X_2 - X_1)$$

$$\frac{d}{dt}[Y_1] = [B][X_1] - [X_1]^2[Y_1] + D_y(Y_2 - Y_1)$$

$$\frac{d}{dt}[X_2] = [A] + [X_2]^2[Y_2] - [B][X_2] - [X_2] + D_x(X_1 - X_2)$$

$$\frac{d}{dt}[Y_2] = [B][X_2] - [X_2]^2[Y_2] + D_y(Y_1 - Y_2)$$

Here, the numerical subscripts indicate which box the material is in. There are additional terms proportional to the diffusion coefficient D that account for the exchange of material between boxes.

If the system is initiated with the same conditions in each box, then a small fluctuation will lead to separation of materials between the two boxes. One box will have a predominance of X, and the other will have a predominance of Y.

Real examples

Real examples of clock reactions are the Belousov-Zhabotinsky reaction (BZ reaction), the Briggs-Rauscher reaction, the Bray-Liebhafsky reaction and the iodine clock reaction. These are oscillatory reactions, and the concentration of products and reactants can be approximated in terms of damped oscillations.

The best-known reaction, the BZ reaction, can be created with a mixture of potassium bromate , malonic acid , and manganese sulfate prepared in a heated solution with sulfuric acid as solvent.

Optics example

Another autocatalytic system is one driven by light coupled to photo-polymerization reactions. In a process termed optical autocatalysis, positive feedback is created between light intensity and photo-polymerization rate, via polymerization-induced increases in the refractive index. Light's preference to occupy regions of higher refractive index results in leakage of light into regions of higher molecular weight, thereby amplifying the photo-chemical reaction. The positive feedback may be expressed as:

Noting that photo-polymerization rate is proportional to intensity and that re-fractive index is proportional to molecular weight, the positive feedback be-tween intensity and photo-polymerization establishes the auto-catalytic behav-ior. Optical auto-catalysis has been shown to result on spontaneous pattern formation in photopolymers. Hosein and co-workers discovered that optical autocatalysis can also occur in photoreactive polymer blends, and that the pro-cess can induce binary phase morphologies with the same pattern as the light profile. The light undergoes optical modulation instability, spontaneous divid-ing into a multitude of optical filaments, and the polymer system thereby forms filaments within the blend structure. The result is a new system that couples optical autocatalytic behavior to spinodal decompositon.

Biological example

It is known that an important metabolic cycle, glycolysis, displays temporal order. Glycolysis consists of the degradation of one molecule of glucose and the overall production of two molecules of ATP. The process is therefore of great importance to the energetics of living cells. The global glycolysis reaction involves glucose, ADP, NAD, pyruvate, ATP, and NADH.

The details of the process are quite involved, however, a section of the process is autocatalyzed by phosphofructokinase (PFK). This portion of the process is responsible for oscillations in the pathway that lead to the process oscillating between an active and an inactive form. Thus, the autocatalytic reaction can modulate the process.

Phase transitions

The initial amounts of reactants determine the distance from chemical equilibrium of the system. The greater the initial concentrations the further the system is from equilibrium. As the initial concentration increases, an abrupt change in order occurs. This abrupt change is known as phase transition. At the phase transition, fluctuations in macroscopic quantities, such as chemical concentrations, increase as the system oscillates between the more ordered state (lower entropy, such as ice) and the more disordered state (higher entropy, such as liquid water). Also, at the phase transition, macroscopic equations, such as the rate equations, fail. Rate equations can be derived from microscopic considerations. The derivations typically rely on a mean field theory approximation to microscopic dynamical equations. Mean field theory breaks down in the presence of large fluctuations (see Mean field theory article for a discussion). Therefore, since large fluctuations occur in the neighborhood of a phase transition, macroscopic equations, such as rate equations, fail. As the initial concentration increases further, the system settles into an ordered state in which fluctuations are again small. (see Prigogine reference)

Asymmetric autocatalysis

Asymmetric autocatalysis occurs when the reaction product is chiral and thus acts as a chiral catalyst for its own production. Reactions of this type, such as the Soai reaction, have the property that they can amplify a very small enantiomeric excess into a large one. This has been proposed as an important step in the origin of biological homochirality.

Role in origin of life

In 1995 Stuart Kauffman proposed that life initially arose as autocatalytic chemical networks.

British ethologist Richard Dawkins wrote about autocatalysis as a potential explanation for abiogenesis in his 2004 book *The Ancestor's Tale*. He cites experiments performed by Julius Rebek and his colleagues at the Scripps Research Institute in California in which they combined amino adenosine and pentafluorophenyl ester with the autocatalyst amino adenosine triacid ester (AATE). One system from the experiment contained variants of AATE which catalysed the synthesis of themselves. This experiment demonstrated the possibility that autocatalysts could exhibit competition within a population of entities with heredity, which could be interpreted as a rudimentary form of natural selection, and that certain environmental changes (such as irradiation) could alter the chemical structure of some of these self-replicating molecules (an analogue for mutation) in such ways that could either boost or interfere with its ability to react, thus boosting or interfering with its ability to replicate and spread in the population.

Autocatalysis plays a major role in the processes of life. Two researchers who have emphasised its role in the origins of life are Robert Ulanowicz[225] and Stuart Kauffman.[226]

Autocatalysis occurs in the initial transcripts of rRNA. The introns are capable of excising themselves by the process of two nucleophilic transesterification reactions. The RNA able to do this is sometimes referred to as a ribozyme. Additionally, the citric acid cycle is an autocatalytic cycle run in reverse.

Ultimately, biological metabolism itself can be seen as a vast autocatalytic set, in that all of the molecular constituents of a biological cell are produced by reactions involving this same set of molecules.

Examples of autocatalytic reactions

- DNA replication
- Haloform reaction
- Formose reaction (also known as Butlerov reaction)
- Tin pest
- Reaction of permanganate with oxalic acid
- Vinegar syndrome
- Binding of oxygen by haemoglobin
- The spontaneous degradation of aspirin into salicylic acid and acetic acid, causing very old aspirin in sealed containers to smell mildly of vinegar.
- The α-bromination of acetophenone with bromine.
- Liesegang rings
- Autocatalytic surface growth of metal nanoparticles in solution phase

External links

- Some Remarks on Autocatalysis and Autopoiesis[227] (Barry McMullin)
- Jain, Sanjay; Krishna, Sandeep (21 December 1998). "Autocatalytic Sets and the Growth of Complexity in an Evolutionary Model". *Physical Review Letters*. **81** (25): 5684–5687. arXiv: adap-org/9809003[228] ⨀ . Bibcode: 1998PhRvL..81.5684J[229]. doi: 10.1103/PhysRevLett.81.5684[230].

Homochirality

Homochirality

Homochirality is a uniformity of chirality, or handedness. Objects are *chiral* when they cannot be superimposed on their mirror images. For example, the left and right hands of a human are approximately mirror images of each other but are not their own mirror images, so they are *chiral*. In biology, the 20 natural amino acids are homochiral, being L-chiral (left-handed), while sugars are D-chiral (right-handed). *Homochirality* can also refer to *enantiomerically pure* substances in which all the constituents are the same enantiomer (a right-handed or left-handed version of an atom or molecule), but some sources discourage this use of the term.

It is unclear whether homochirality has a purpose, however, it appears to be a form of information storage. One suggestion is that it reduces entropy barriers in the formation of large organized molecules. It has been experimentally verified that amino acids form large aggregates in larger abundance from enantiopure substrates than from racemic ones.Wikipedia:Citation needed

It is not clear whether homochirality emerged before or after life, and many mechanisms for its origin have been proposed. Some of these models propose three distinct steps: *mirror-symmetry breaking* creates a minute enantiomeric imbalance, *chiral amplification* builds on this imbalance, and *chiral transmission* is the transfer of chirality from one set of molecules to another.

In biology

Amino acids are the building blocks of peptides and enzymes while sugar-peptide chains are the backbone of RNA and DNA. In biological organisms, amino acids appear almost exclusively in the left-handed form (L-amino acids) and sugars in the right-handed form (R-sugars). Since the enzymes catalyze

113

reactions, they enforce homochirality on a great variety of other chemicals, including hormones, toxins, fragrances and food flavors.[493-494]

Biological organisms easily discriminate between molecules with different chiralities. This can affect physiological reactions such as smell and taste. Carvone, a terpenoid found in essential oils, smells like mint in its L-form and caraway in its R-form.[494] Limonene tastes like lemons when right-handed and oranges when left-handed.[168]

Homochirality also affects the response to drugs. Thalidomide, in its left-handed form, cures morning sickness; in its right-handed form, it causes birth defects.[168] Unfortunately, even if a pure left-handed version is administered, some of it can convert to the right-handed form in the patient. Many drugs are available as both a racemic mixture (equal amounts of both chiralities) and an enantiopure drug (only one chirality). Depending on the manufacturing process, enantiopure forms can be more expensive to produce than stereochemical mixtures.[168]

Chiral preferences can also be found at a macroscopic level. Snail shells can be right-turning or left-turning helices, but one form or the other is strongly preferred in a given species. In the edible snail *Helix pomatia*, only one out of 20,000 is left-helical. The coiling of plants can have a preferred chirality and even the chewing motion of cows has a 10% excess in one direction.

Origins

Symmetry breaking

Known mechanisms for the production of non-racemic mixtures from racemic starting materials include: asymmetric physical laws, such as the electroweak interaction; asymmetric environments, such as those caused by circularly polarized light, quartz crystals, or the Earth's rotation; and statistical fluctuations during racemic synthesis.[231] Once established, chirality would be selected for. A small enantiomeric excess can be amplified into a large one by asymmetric autocatalysis, such as in the Soai reaction. In asymmetric autocatalysis, the catalyst is a chiral molecule, which means that a chiral molecule is catalysing its own production. An initial enantiomeric excess, such as can be produced by polarized light, then allows the more abundant enantiomer to outcompete the other.

One supposition is that the discovery of an enantiomeric imbalance in molecules in the Murchison meteorite supports an extraterrestrial origin of homochirality: there is evidence for the existence of circularly polarized light originating from Mie scattering on aligned interstellar dust particles which may trigger the formation of an enantiomeric excess within chiral material in space.

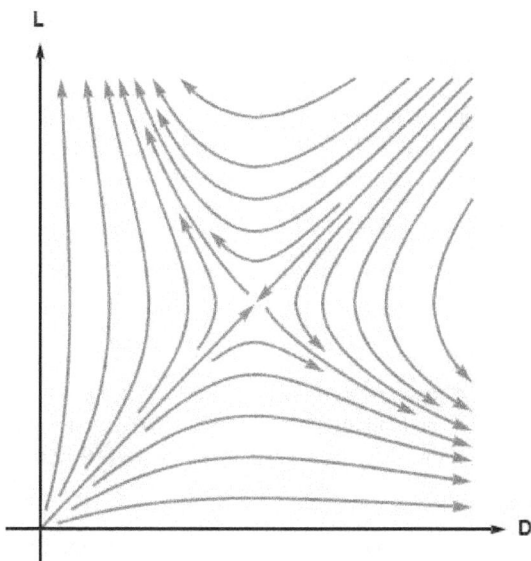

Figure 48: *Phase portrait of Frank's model: starting from almost everywhere in L-D plane (except L = D line), the system approaches to one of the homochiral states*

Interstellar and near-stellar magnetic fields can align dust particles in this fashion. Another speculation (the Vester-Ulbricht hypothesis) suggests that fundamental chirality of physical processes such as that of the beta decay (see Parity violation) leads to slightly different half-lives of biologically relevant molecules. Homochirality may also result from spontaneous absolute asymmetric synthesis.

It is also possible that homochirality is simply a result of the natural autoamplification process of life—that either the formation of life as preferring one chirality or the other was a chance rare event which happened to occur with the chiralities we observe, or that all chiralities of life emerged rapidly but due to catastrophic events and strong competition, the other unobserved chiral preferences were wiped out by the preponderance and metabolic, enantiomeric enrichment from the 'winning' chirality choices.Wikipedia:Citation needed The emergence of chirality consensus as a natural autoamplification process has been associated with the 2nd law of thermodynamics.

Amplification

Theory

In 1953, Charles Frank proposed a model to demonstrate that homochirality is a consequence of autocatalysis. In his model the L and D enantiomers of a chiral molecule are autocatalytically produced from an achiral molecule A

$$A + L \rightarrow 2L, \ A + D \rightarrow 2D$$

while suppressing each other through a reaction that he called *mutual antagonism*

$$L + D \rightarrow \emptyset.$$

In this model the racemic state is unstable in the sense that the slightest enantiomeric excess will be amplified to a completely homochiral state. This can be shown by computing the reaction rates from the law of mass action:

where k_a is the rate constant for the autocatalytic reactions, k_d is the rate constant for mutual antagonism reaction, and the concentration of A is kept constant for simplicity. By defining the enantiomeric excess Φ as

we can compute the rate of change of enatiomeric excess using chain rule from the rate of change of the concentrations of enantiomeres L and D.

Linear stability analysis of this equation shows that the racemic state $\Phi = 0$ is unstable. Starting from almost everywhere in the concentration space, the system evolves to a homochiral state.

It is generally understood that autocatalysis alone does not yield to homochirality, and the presence of the mutually antagonistic relationship between the two enantiomers is necessary for the instability of the racemic mixture. However, recent studies show that homochirality could be achieved from autocatalysis in the absence of the mutually antagonistic relationship, but the underlying mechanism for symmetry-breaking is different.

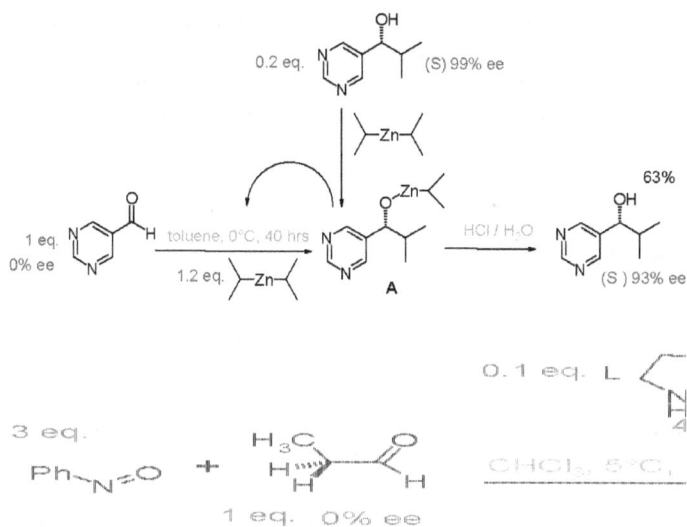

Experiments

Laboratory experiments exist demonstrating how in certain autocatalytic reaction systems the presence of a small amount of reaction product with enantiomeric excess at the start of the reaction can result in a much larger enantiomeric excess at the end of the reaction. In the Soai reaction, pyrimidine-5-
carbaldehyde (*Scheme 1*) is alkylated by diisopropylzinc to the corresponding
pyrimidyl alcohol. Because the initial reaction product is also an effective catalyst the reaction is autocatalytic. The presence of just 0.2 equivalent of the
alcohol S-enantiomer at the start of the reaction is sufficient to amplify the
enantiomeric excess to 93%.

Another study concerns the proline catalyzed aminoxylation of propionalde-
hyde by nitrosobenzene (*scheme 2*). In this system too the presence of enantioenriched catalyst drives the reaction towards one of the two possible optical
isomers.

Serine octamer clusters are also contenders. These clusters of 8 serine
molecules appear in mass spectrometry with an unusual homochiral preference, however there is no evidence that such clusters exist under non-ionizing
conditions and amino acid phase behavior is far more prebiotically relevant.
The recent observation that partial sublimation of a 10% enantioenriched sample of leucine results in up to 82% enrichment in the sublimate shows that

enantioenrichment of amino acids could occur in space. Partial sublimation processes can take place on the surface of meteors where large variations in temperature exist. This finding may have consequences for the development of the Mars Organic Detector scheduled for launch in 2013 which aims to recover trace amounts of amino acids from the Mars surface exactly by a sublimation technique.

A high asymmetric amplification of the enantiomeric excess of sugars are also present in the amino acid catalyzed asymmetric formation of carbohydrates

One classic study involves an experiment that takes place in the laboratory. When sodium chlorate is allowed to crystallize from water and the collected crystals examined in a polarimeter, each crystal turns out to be chiral and either the L form or the D form. In an ordinary experiment the amount of L crystals collected equals the amount of D crystals (corrected for statistical effects). However, when the sodium chlorate solution is stirred during the crystallization process the crystals are either exclusively L or exclusively D. In 32 consecutive crystallization experiments 14 experiments deliver D-crystals and 18 others L-crystals. The explanation for this symmetry breaking is unclear but is related to autocatalysis taking place in the nucleation process.

In a related experiment, a crystal suspension of a racemic amino acid derivative continuously stirred, results in a 100% crystal phase of one of the enantiomers because the enantiomeric pair is able to equilibrate in solution (compare with dynamic kinetic resolution).

Transmission

Many strategies in asymmetric synthesis are built on chiral transmission. Especially important is the so-called organocatalysis of organic reactions by proline for example in Mannich reactions.

Optical resolution in racemic amino acids

There exists no theory elucidating correlations among L-amino acids. If one takes, for example, alanine, which has a small methyl group, and phenylalanine, which has a larger benzyl group, a simple question is in what aspect, L-alanine resembles L-phenylalanine more than D-phenylalanine, and what kind of mechanism causes the selection of all L-amino acids. Because it might be possible that alanine was L and phenylalanine was D.

It was reported in 2004 that excess racemic D,L-asparagine (Asn), which spontaneously forms crystals of either isomer during recrystallization, induces asymmetric resolution of a co-existing racemic amino acid such as arginine

(Arg), aspartic acid (Asp), glutamine (Gln), histidine (His), leucine (Leu), methionine (Met), phenylalanine (Phe), serine (Ser), valine (Val), tyrosine (Tyr), and tryptophan (Trp). The enantiomeric excess ee = $100 \times (L-D)/(L+D)$ of these amino acids was correlated almost linearly with that of the inducer, i.e., Asn. When recrystallizations from a mixture of 12 D,L-amino acids (Ala, Asp, Arg, Glu, Gln, His, Leu, Met, Ser, Val, Phe, and Tyr) and excess D,L-Asn were made, all amino acids with the same configuration with Asn were preferentially co-crystallized. It was incidental whether the enrichment took place in L- or D-Asn, however, once the selection was made, the co-existing amino acid with the same configuration at the α-carbon was preferentially involved because of thermodynamic stability in the crystal formation. The maximal ee was reported to be 100%. Based on these results, it is proposed that a mixture of racemic amino acids causes spontaneous and effective optical resolution, even if asymmetric synthesis of a single amino acid does not occur without an aid of an optically active molecule.

This is the first study elucidating reasonably the formation of chirality from racemic amino acids with experimental evidences.

History of term

This term was introduced by Kelvin in 1904, the year that he published his Baltimore Lecture of 1884. Kelvin used the term homochirality as a relationship between two molecules, i.e. two molecule are homochiral if they have the same chirality. Recently, however, homochiral has been used in the same sense as enantiomerically pure. This is permitted in some journals (but not encouraged),[342][232] its meaning changing into the preference of a process or system for a single optical isomer in a pair of isomers in these journals.

Further reading

<templatestyles src="Template:Refbegin/styles.css" />

- Bailey, Jeremy (28 August 1998). "Stellar Circular Polarization and Biomolecular Homochirality"[233]. *ScienceWeek*. Archived from the original[234] on 30 November 2010. Retrieved 5 May 2018.
- Blackmond, Donna G.; Miller, Rom (21 June 2004). "How left-handed amino acids got ahead: a demonstration of the evolution of biological homochirality in the lab"[235]. *www.imperial.ac.uk* (Press release). Imperial College London. Retrieved 5 May 2018.
- Hegstrom, Roger A.; Kondepudi, Dilip K. (1990). "The Handedness of the Universe". *Scientific American*. **262** (1): 108–115. JSTOR 24996649[236].

- Jafarpour, Farshid; Biancalani, Tommaso; Goldenfeld, Nigel (10 March 2017). "Noise-induced symmetry breaking far from equilibrium and the emergence of biological homochirality". *Physical Review E*. **95** (3). Bibcode: 2017PhRvE..95c2407J[237]. doi: 10.1103/PhysRevE.95.032407[238].
- Rouhi, A. Maureen (17 June 2004). "On the Genesis of Homochirality"[239]. *Chemical & Engineering News*. American Chemical Society. Retrieved 5 May 2018.

External links

- Observations Support Homochirality Theory[240]. Photonics Technology-World November 1998.
- Origins of Homochirality[241]. Conference in Nordita Stockholm, February 2008.

Protocells

Protocell

A **protocell** (or **protobiont**) is a self-organized, endogenously ordered, spherical collection of lipids proposed as a stepping-stone to the origin of life. A central question in evolution is how simple protocells first arose and how they could differ in reproductive output, thus enabling the accumulation of novel biological emergences over time, i.e. biological evolution. Although a functional protocell has not yet been achieved in a laboratory setting, the goal to understand the process appears well within reach.

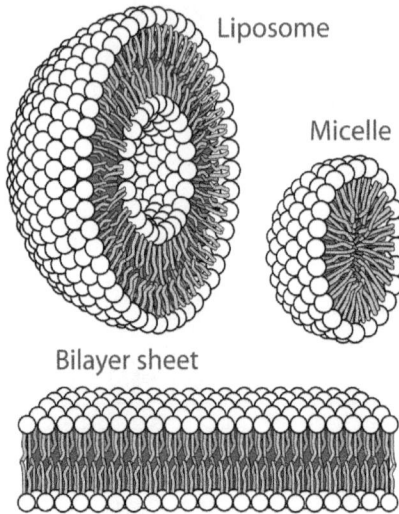

Figure 49: *The three main structures phospholipids form in solution; the liposome (a closed bilayer), the micelle and the bilayer.*

Overview

Compartmentalization was important in the origins of life.Wikipedia:Citation needed Membranes form enclosed compartments that are separate from the external environment, thus providing the cell with functionally specialized aqueous spaces. Because lipid bilayer of membranes is impermeable to most hydrophilic molecules (dissolved by water), cells have membrane transport systems that achive the import of nutritive molecules as well as the export of waste.Wikipedia:Verifiability It is very challenging to construct protocells from molecular assemblies. An important step in this challenge is the achievement of vesicle dynamics that are relevant to cellular functions, such as membrane trafficking and self-reproduction, using amphiphilic molecules. On the primitive Earth, numerous chemical reactions of organic compounds produced the ingredients of life.Wikipedia:Citation needed Of these substances, amphiphilic molecules might be the first player in the evolution from molecular assembly to cellular life.[243,244] A step from vesicle toward protocell might be to develop self-reproducing vesicles coupled with the metabolic system.

Selectivity for compartmentalization

Self-assembled vesicles are essential components of primitive cells. The second law of thermodynamics requires that the universe move in a direction in which disorder (or entropy) increases, yet life is distinguished by its great degree of organization. Therefore, a boundary is needed to separate life processes from non-living matter. The cell membrane is the only cellular structure that is found in all of the cells of all of the organisms on Earth.

Researchers Irene A. Chen and Jack W. Szostak (Nobel Prize in Physiology or Medicine 2009) amongst others, demonstrated that simple physicochemical properties of elementary protocells can give rise to essential cellular behaviors, including primitive forms of Darwinian competition and energy storage. Such cooperative interactions between the membrane and encapsulated contents could greatly simplify the transition from replicating molecules to true cells. Competition for membrane molecules would favor stabilized membranes, suggesting a selective advantage for the evolution of cross-linked fatty acids and even the phospholipids of today. This micro-encapsulation allowed for metabolism within the membrane, exchange of small molecules and prevention of passage of large substances across it. The main advantages of encapsulation include increased solubility of the cargo and creating energy in the form of chemical gradient. Energy is thus often said to be stored by cells in the structures of molecules of substances such as carbohydrates (including sugars), lipids, and proteins, which release energy when chemically combined with oxygen during cellular respiration.

Energy gradient

A March 2014 study by NASA's Jet Propulsion Laboratory demonstrated a unique way to study the origins of life: fuel cells. Fuel cells are similar to biological cells in that electrons are also transferred to and from molecules. In both cases, this results in electricity and power. The study states that one important factor was that the Earth provides electrical energy at the seafloor. "This energy could have kick-started life and could have sustained life after it arose. Now, we have a way of testing different materials and environments that could have helped life arise not just on Earth, but possibly on Mars, Europa and other places in the Solar System."

Vesicles and micelles

When phospholipids are placed in water, the molecules spontaneously arrange such that the tails are shielded from the water, resulting in the formation of membrane structures such as bilayers, vesicles, and micelles. In modern cells,

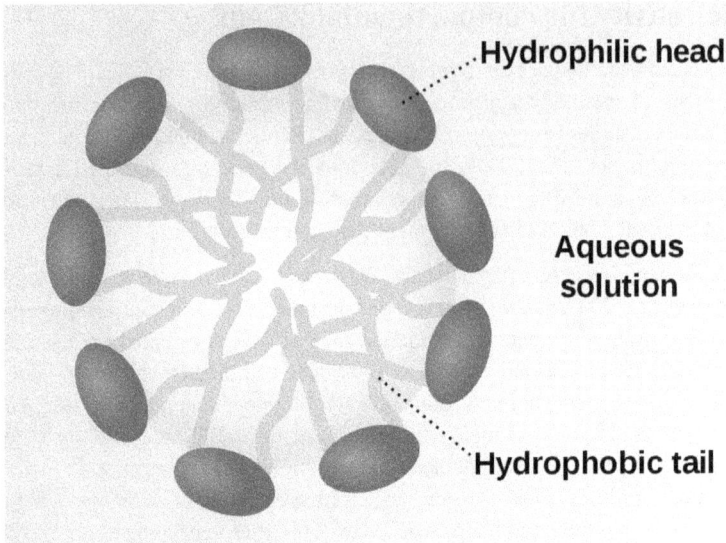

Figure 50: *Scheme of a micelle spontaneously formed by phospholipids in an aqueous solution*

vesicles are involved in metabolism, transport, buoyancy control, and enzyme storage. They can also act as natural chemical reaction chambers. A typical vesicle or micelle in aqueous solution forms an aggregate with the hydrophilic "head" regions in contact with surrounding solvent, sequestering the hydrophobic single-tail regions in the micelle centre. This phase is caused by the packing behavior of single-tail lipids in a bilayer. Although the protocellular self-assembly process that spontaneously form lipid *monolayer* vesicles and micelles in nature resemble the kinds of primordial vesicles or protocells that might have existed at the beginning of evolution, they are not as sophisticated as the *bilayer* membranes of today's living organisms.

Rather than being made up of phospholipids, however, early membranes may have formed from monolayers or bilayers of fatty acids, which may have formed more readily in a prebiotic environment. Fatty acids have been synthesized in laboratories under a variety of prebiotic conditions and have been found on meteorites, suggesting their natural synthesis in nature.

Oleic acid vesicles represent good models of membrane protocells that could have existed in prebiotic times.

Electrostatic interactions induced by short, positively charged, hydrophobic peptides containing 7 amino acids in length or fewer, can attach RNA to a vesicle membrane, the basic cell membrane.

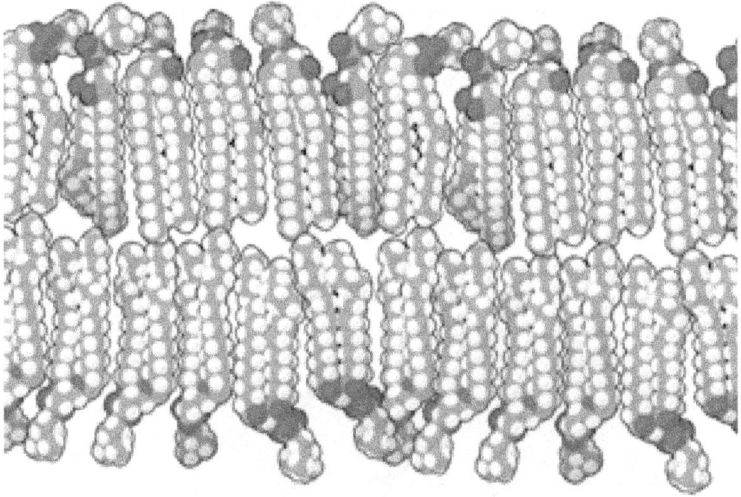

Figure 51: *This fluid lipid bilayer cross section is made up entirely of phosphatidylcholine.*

Geothermal ponds and clay

Scientists have come to conclude that life began in hydrothermal vents in the
deep sea, but a 2012 study suggests that inland pools of condensed and cooled
geothermal vapor have the ideal characteristics for the origin of life. The conclusion is based mainly on the chemistry of modern cells, where the cyto-
plasm is rich in potassium, zinc, manganese, and phosphate ions, which are
not widespread in marine environments. Such conditions, the researchers argue, are found only where hot hydrothermal fluid brings the ions to the surface—places such as geysers, mud pots, fumaroles and other geothermal features. Within these fuming and bubbling basins, water laden with zinc and
manganese ions could have collected, cooled and condensed in shallow pools.

Another study in the 1990s showed that montmorillonite clay can help create
RNA chains of as many as 50 nucleotides joined together spontaneously into a
single RNA molecule. Later, in 2002, it was discovered that by adding montmorillonite to a solution of fatty acid micelles (lipid spheres), the clay sped up
the rate of vesicle formation 100-fold.

Research has shown that some minerals can catalyze the stepwise formation of hydrocarbon tails of fatty acids from hydrogen and carbon monoxide
gases—gases that may have been released from hydrothermal vents or geysers.
Fatty acids of various lengths are eventually released into the surrounding water, but vesicle formation requires a higher concentration of fatty acids, so it

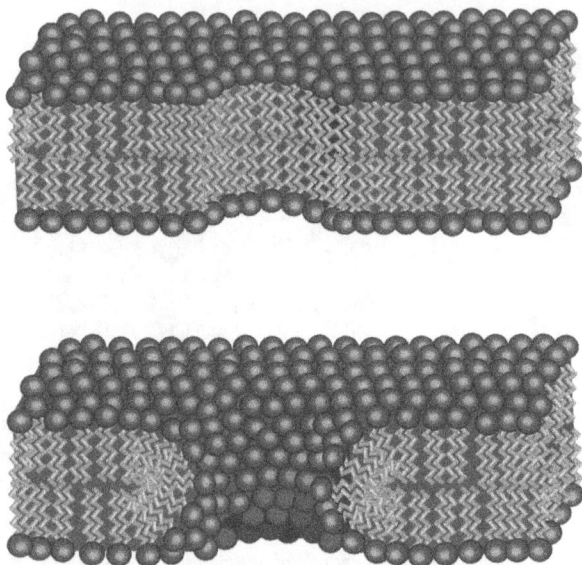

Figure 52: *Schematic showing two possible conformations of the lipids at the edge of a pore. In the top image the lipids have not rearranged, so the pore wall is hydrophobic. In the bottom image some of the lipid heads have bent over, so the pore wall is hydrophilic.*

is suggested that protocell formation started at land-bound hydrothermal vents such as geysers, mud pots, fumaroles and other geothermal features where water evaporates and concentrates the solute.

Montmorillonite bubbles

Another group suggests that primitive cells might have formed inside inorganic clay microcompartments, which can provide an ideal container for the synthesis and compartmentalization of complex organic molecules. Clay-armored *bubbles* form naturally when particles of montmorillonite clay collect on the outer surface of air bubbles under water. This creates a semi permeable vesicle from materials that are readily available in the environment. The authors remark that montmorillonite is known to serve as a chemical catalyst, encouraging lipids to form membranes and single nucleotides to join into strands of RNA. Primitive reproduction can be envisioned when the clay bubbles burst, releasing the lipid membrane-bound product into the surrounding medium.

Membrane transport

Instead of the more popular phospholipids of modern cells, the membrane of protocells in the RNA world would be composed of fatty acids, and that such membranes have relatively high permeability to ions and small molecules, such as nucleoside monophosphate (NMP), nucleoside diphosphate (NDP), and nucleoside triphosphate (NTP), and may withstand millimolar concentrations of Mg^{2+}. Osmotic pressure also plays a significant role in protocell membrane transport.

It has been proposed that electroporation resulting from lightning strikes could be a mechanism of natural horizontal gene transfer. Electroporation is the rapid increase in bilayer permeability induced by the application of a large artificial electric field across the membrane. During electroporation in laboratory procedures, the lipid molecules are not chemically altered but simply shift position, opening up a pore (hole) that acts as the conductive pathway through the bilayer as it is filled with water. The mechanism is the creation of nanometer sized water-filled holes in the membrane. Experimentally, electroporation is used to introduce hydrophilic molecules into cells. It is a particularly useful technique for large highly charged molecules such as DNA and RNA, which would never passively diffuse across the hydrophobic bilayer core. Because of this, electroporation is one of the key methods of transfection as well as bacterial transformation.

Fusion

Some molecules or particles are too large or too hydrophilic to pass through a lipid bilayer, but can be moved across the cell membrane through fusion or budding of vesicles. This may have eventually led to mechanisms that facilitate movement of molecules to the inside (endocytosis) or to release its contents into the extracellular space (exocytosis).

Artificial models

Langmuir-Blodgett deposition

Starting with a technique commonly used to deposit molecules on a solid surface, Langmuir–Blodgett deposition, scientists are able to assemble phospholipid membranes of arbitrary complexity layer by layer. These artificial phospholipid membranes support functional insertion both of purified and of *in situ* expressed membrane proteins. The technique could help astrobiologists understand how the first living cells originated.

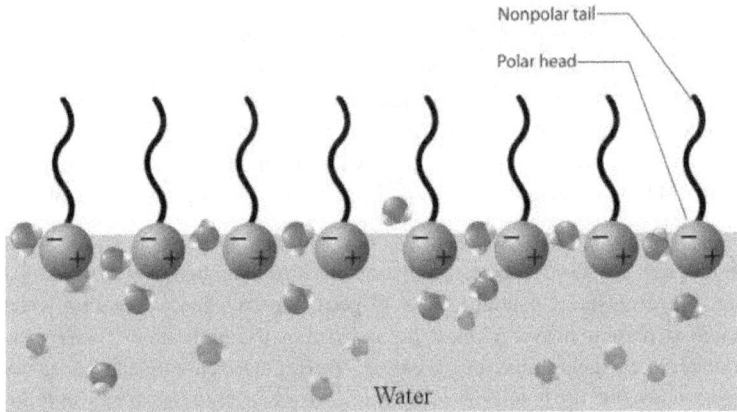

Figure 53: *Surfactant molecules arranged on an air – water interface*

Jeewanu protocells

Jeewanu protocells are synthetic chemical particles that possess cell-like structure and seem to have some functional living properties. First synthesized in 1963 from simple minerals and basic organics while exposed to sunlight, it is still reported to have some metabolic capabilities, the presence of semipermeable membrane, amino acids, phospholipids, carbohydrates and RNA-like molecules. However, the nature and properties of the Jeewanu remains to be clarified.

In a similar synthesis experiment a frozen mixture of water, methanol, ammonia and carbon monoxide was exposed to ultraviolet (UV) radiation. This combination yielded large amounts of organic material that self-organised to form globules or vesicles when immersed in water. The investigating scientist considered these globules to resemble cell membranes that enclose and concentrate the chemistry of life, separating their interior from the outside world. The globules were between 10 to 40 micrometres (0.00039 to 0.00157 in), or about the size of red blood cells. Remarkably, the globules fluoresced, or glowed, when exposed to UV light. Absorbing UV and converting it into visible light in this way was considered one possible way of providing energy to a primitive cell. If such globules played a role in the origin of life, the fluorescence could have been a precursor to primitive photosynthesis. Such fluorescence also provides the benefit of acting as a sunscreen, diffusing any damage that otherwise would be inflicted by UV radiation. Such a protective function would have been vital for life on the early Earth, since the ozone layer, which blocks out the sun's most destructive UV rays, did not form until after photosynthetic life began to produce oxygen.

Ethics and controversy

Protocell research has created controversy and opposing opinions, including critics of the vague definition of "artificial life". The creation of a basic unit of life is the most pressing ethical concern, although the most widespread worry about protocells is their potential threat to human health and the environment through uncontrolled replication.

External links

- "Protocells: Bridging Nonliving and Living Matter." Edited by Steen Rasmussen, Mark A. Bedau, Liaochai Chen, David Deamer, David Krakauer, Norman, H.Packard and Peter F. Stadler. MIT Press, Cambridge, Massachusetts. 2008.
- " Living Chemistry & A Natural History of Protocells.[245]" Synth-ethic: Art and Synthetic Biology Exhibition (2013) at the Natural History Museum, Vienna, Austria.

RNA world

RNA world

The **RNA world** is a hypothetical stage in the evolutionary history of life on Earth, in which self-replicating RNA molecules proliferated before the evolution of DNA and proteins. The term also refers to the hypothesis that posits the existence of this stage.

Alexander Rich first proposed the concept of the RNA world in 1962, and Walter Gilbert coined the term in 1986. Alternative chemical paths to life have been proposed, and RNA-based life may not have been the first life to exist. Even so, the evidence for an RNA world is strong enough that the hypothesis has gained wide acceptance.

Like DNA, RNA can store and replicate genetic information; like protein enzymes, RNA enzymes (ribozymes) can catalyze (start or accelerate) chemical reactions that are critical for life. One of the most critical components of cells, the ribosome, is composed primarily of RNA. Ribonucleotide moieties in many coenzymes, such as Acetyl-CoA, NADH, FADH and F420, have long been thought ofWikipedia:Manual of Style/Words to watch#Unsupported attributions as surviving remnants of covalently bound coenzymes in an RNA world.

Although RNA is fragile, some ancient RNAs may have evolved the ability to methylate other RNAs to protect them.

If the RNA world existed, it was probably followed by an age characterized by the evolution of ribonucleoproteins (RNP world), which in turn ushered in the era of DNA and longer proteins. DNA has better stability and durability than RNA; this may explain why it became the predominant storage molecule. Protein enzymes may have come to replace RNA-based ribozymes as biocatalysts because their greater abundance and diversity of monomers makes them more versatile. As some co-factors contain both nucleotide and amino-acid characteristics, it may be that amino acids, peptides and finally proteins initially were co-factors for ribozymes.

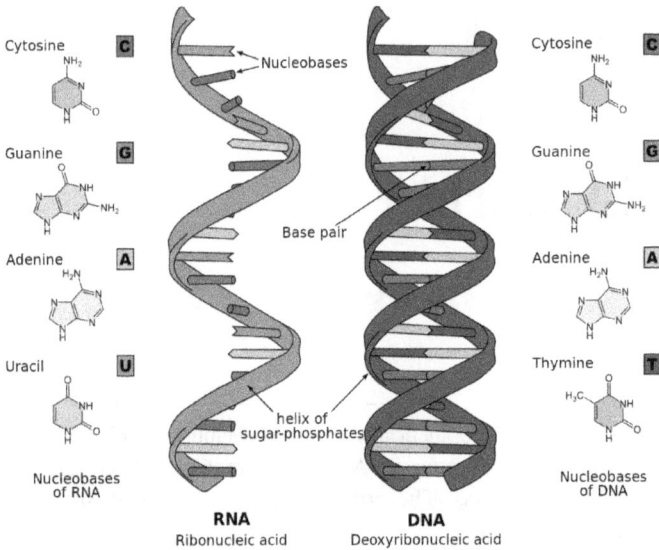

Figure 54: *A comparison of RNA (left) with DNA (right),
showing the helices and nucleobases each employs*

History

One of the challenges in studying abiogenesis is that the system of reproduction and metabolism utilized by all extant life involves three distinct types of inter-dependent macromolecules (DNA, RNA, and protein). This suggests that life could not have arisen in its current form, which has led researchers to hypothesize mechanisms whereby the current system might have arisen from a simpler precursor system. The concept of RNA as a primordial molecule can be found in papers by Francis Crick and Leslie Orgel, as well as in Carl Woese's 1967 book *The Genetic Code*.[246] In 1962, the molecular biologist Alexander Rich posited much the same idea in an article he contributed to a volume issued in honor of Nobel-laureate physiologist Albert Szent-Györgyi. Hans Kuhn in 1972 laid out a possible process by which the modern genetic system might have arisen from a nucleotide-based precursor, and this led Harold White in 1976 to observe that many of the cofactors essential for enzymatic function are either nucleotides or could have been derived from nucleotides. He proposed that these nucleotide cofactors represent "fossils of nucleic acid enzymes". The phrase "RNA World" was first used by Nobel laureate Walter Gilbert in 1986, in a commentary on how recent observations of the catalytic properties of various forms of RNA fit with this hypothesis.

Properties of RNA

The properties of RNA make the idea of the RNA world hypothesis conceptually plausible, though its general acceptance as an explanation for the origin of life requires further evidence. RNA is known to form efficient catalysts and its similarity to DNA makes clear its ability to store information. Opinions differ, however, as to whether RNA constituted the first autonomous self-replicating system or was a derivative of a still-earlier system. One version of the hypothesis is that a different type of nucleic acid, termed *pre-RNA*, was the first one to emerge as a self-reproducing molecule, to be replaced by RNA only later. On the other hand, the discovery in 2009 that activated pyrimidine ribonucleotides can be synthesized under plausible prebiotic conditions suggests that it is premature to dismiss the RNA-first scenarios. Suggestions for 'simple' *pre-RNA* nucleic acids have included peptide nucleic acid (PNA), threose nucleic acid (TNA) or glycol nucleic acid (GNA). Despite their structural simplicity and possession of properties comparable with RNA, the chemically plausible generation of "simpler" nucleic acids under prebiotic conditions has yet to be demonstrated.

RNA as an enzyme

RNA enzymes, or ribozymes, are found in today's DNA-based life and could be examples of living fossils. Ribozymes play vital roles, such as that of the ribosome, an RNA-protein complex responsible for protein synthesis. Many other ribozyme functions exist; for example, the hammerhead ribozyme performs self-cleavage and an RNA polymerase ribozyme can synthesize a short RNA strand from a primed RNA template.

Among the enzymatic properties important for the beginning of life are:

Self-replication

The ability to self-replicate, or synthesize other RNA molecules; relatively short RNA molecules that can synthesize others have been artificially produced in the lab. The shortest was 165-bases long, though it has been estimated that only part of the molecule was crucial for this function. One version, 189-bases long, had an error rate of just 1.1% per nucleotide when synthesizing an 11 nucleotide long RNA strand from primed template strands. This 189 base pair ribozyme could polymerize a template of at most 14 nucleotides in length, which is too short for self replication, but is a potential lead for further investigation. The longest primer extension performed by a ribozyme polymerase was 20 bases.[247] In 2016, researchers reported the use of in vitro evolution to improve dramatically the activity and generality of an RNA polymerase ribozyme by selecting variants that can synthesize functional RNA molecules from an RNA template. Each RNA polymerase ribozyme was engineered to remain linked to its new, synthesized RNA strand, this allowed the team to isolate successful polymerases. The isolated RNA polymerases were again used for another round of evolution. After several rounds of evolution, they obtained one RNA polymerase ribozyme called 24-3 that was able to copy almost any other RNA, from small catalysts to long RNA based enzymes. Particular RNAs were amplified up to 10,000 times, a first RNA version of the polymerase chain reaction (PCR). The RNA polymerase is not yet able to make copies of itself.

Catalysis
The ability to catalyze simple chemical reactions—which would enhance creation of molecules that are building blocks of RNA molecules (i.e., a strand of RNA that would make creating more strands of RNA easier). Relatively short RNA molecules with such abilities have been artificially formed in the lab. A recent study showed that almost any nucleic acid can evolve into a catalytic sequence under appropriate selection. For instance, an arbitrarily chosen 50-nucleotide DNA fragment encoding for the *Bos taurus* (cattle) albumin mRNA was subjected to test-tube evolution to derive a catalytic DNA (Deoxyribozyme, also called DNAzyme) with RNA-cleavage activity. After only a few weeks, a DNAzyme with significant catalytic activity had evolved. In general, DNA is much more chemically inert than RNA and hence much more resistant to obtaining catalytic properties. If in vitro evolution works for DNA it will happen much more easily with RNA.

Amino acid-RNA ligation
The ability to conjugate an amino acid to the 3'-end of an RNA in order to use its chemical groups or provide a long-branched aliphatic side-chain.

Peptide bond formation
The ability to catalyse the formation of peptide bonds between amino acids to produce short peptides or longer proteins. This is done in modern cells by ribosomes, a complex of several RNA molecules known as rRNA together with many proteins. The rRNA molecules are thought responsible for its enzymatic activity, as no amino acid molecules lie within 18Å of the enzyme's active site, and, when the majority of the amino acids in the ribosome were stringently removed, the resulting ribosome retained its full peptidyl transferase activity, fully able to catalyze the formation of peptide bonds between amino acids. A much shorter RNA molecule has been synthesized in the laboratory with the ability to form peptide bonds, and it has been suggested that rRNA has evolved from a similar molecule. It has also been suggested that amino acids may have initially been involved

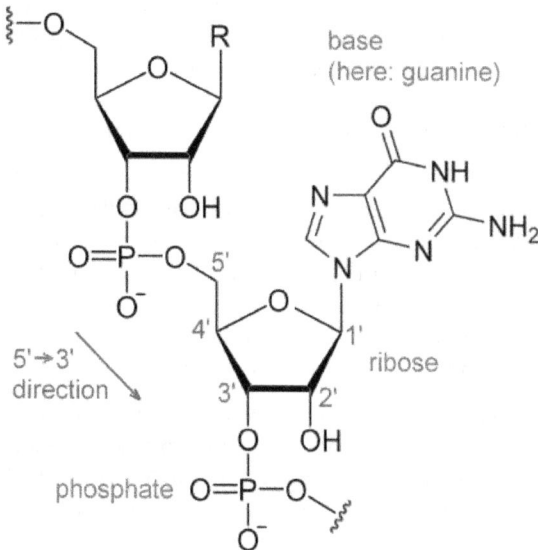

Figure 55: *The major difference between RNA and DNA is the presence of a hydroxyl group at the 2'-position.*

with RNA molecules as cofactors enhancing or diversifying their enzymatic capabilities, before evolving to more complex peptides. Similarly, tRNA is suggested to have evolved from RNA molecules that began to catalyze amino acid transfer.

RNA in information storage

RNA is a very similar molecule to DNA, with only two major chemical differences (the backbone of RNA uses ribose instead of deoxyribose and its nucleobases include uracil instead of thymine). The overall structure of RNA and DNA are immensely similar—one strand of DNA and one of RNA can bind to form a double helical structure. This makes the storage of information in RNA possible in a very similar way to the storage of information in DNA. However, RNA is less stable, being more prone to hydrolysis due to the presence of a hydroxyl group at the ribose 2' position.

Comparison of DNA and RNA structure

The major difference between RNA and DNA is the presence of a hydroxyl group at the 2'-position of the ribose sugar in RNA (illustration, right). This group makes the molecule less stable because, when not constrained in a double helix, the 2' hydroxyl can chemically attack the adjacent phosphodiester bond to cleave the phosphodiester backbone. The hydroxyl group also forces the ribose into the C3'-*endo* sugar conformation unlike the C2'-*endo* conformation of the deoxyribose sugar in DNA. This forces an RNA double helix to change from a B-DNA structure to one more closely resembling A-DNA.

RNA also uses a different set of bases than DNA—adenine, guanine, cytosine and uracil, instead of adenine, guanine, cytosine and thymine. Chemically, uracil is similar to thymine, differing only by a methyl group, and its production requires less energy. In terms of base pairing, this has no effect. Adenine readily binds uracil or thymine. Uracil is, however, one product of damage to cytosine that makes RNA particularly susceptible to mutations that can replace a **GC** base pair with a **GU** (wobble) or **AU** base pair.

RNA is thought to have preceded DNA, because of their ordering in the biosynthetic pathways. The deoxyribonucleotides used to make DNA are made from ribonucleotides, the building blocks of RNA, by removing the 2'-hydroxyl group. As a consequence a cell must have the ability to make RNA before it can make DNA.

Limitations of information storage in RNA

The chemical properties of RNA make large RNA molecules inherently fragile, and they can easily be broken down into their constituent nucleotides through hydrolysis. These limitations do not make use of RNA as an information storage system impossible, simply energy intensive (to repair or replace damaged RNA molecules) and prone to mutation. While this makes it unsuitable for current 'DNA optimised' life, it may have been acceptable for more primitive life.

RNA as a regulator

Riboswitches have been found to act as regulators of gene expression, particularly in bacteria, but also in plants and archaea. Riboswitches alter their secondary structure in response to the binding of a metabolite. This change in structure can result in the formation or disruption of a terminator, truncating or permitting transcription respectively. Alternatively, riboswitches may bind or occlude the Shine-Dalgarno sequence, affecting translation. It has been suggested that these originated in an RNA-based world. In addition, RNA thermometers regulate gene expression in response to temperature changes.

Support and difficulties

The RNA world hypothesis is supported by RNA's ability to store, transmit, and duplicate genetic information, as DNA does. RNA can act as a ribozyme, a special type of enzyme. Because it can perform the tasks of both DNA and enzymes, RNA is believed to have once been capable of supporting independent life forms. Some viruses use RNA as their genetic material, rather than DNA.[248] Further, while nucleotides were not found in experiments based on Miller-Urey experiment, their formation in prebiotically plausible conditions was reported in 2009; the purine base known as adenine is merely a pentamer of hydrogen cyanide. Experiments with basic ribozymes, like Bacteriophage Qβ RNA, have shown that simple self-replicating RNA structures can withstand even strong selective pressures (e.g., opposite-chirality chain terminators).[249]

Since there were no known chemical pathways for the abiogenic synthesis of nucleotides from pyrimidine nucleobases cytosine and uracil under prebiotic conditions, it is thought by some that nucleic acids did not contain these nucleobases seen in life's nucleic acids. The nucleoside cytosine has a half-life in isolation of 19 days at 100 °C (212 °F) and 17,000 years in freezing water, which some argue is too short on the geologic time scale for accumulation. Others have questioned whether ribose and other backbone sugars could be stable enough to find in the original genetic material, and have raised the issue that all ribose molecules would have had to be the same enantiomer, as any nucleotide of the wrong chirality acts as a chain terminator.

Pyrimidine ribonucleosides and their respective nucleotides have been prebiotically synthesised by a sequence of reactions that by-pass free sugars and assemble in a stepwise fashion by including nitrogenous and oxygenous chemistries. In a series of publications, John Sutherland and his team at the School of Chemistry, University of Manchester, have demonstrated high yielding routes to cytidine and uridine ribonucleotides built from small 2 and 3 carbon fragments such as glycolaldehyde, glyceraldehyde or glyceraldehyde-3-phosphate, cyanamide and cyanoacetylene. One of the steps in this sequence allows the isolation of enantiopure ribose aminooxazoline if the enantiomeric excess of glyceraldehyde is 60% or greater, of possible interest towards biological homochirality.[250] This can be viewed as a prebiotic purification step, where the said compound spontaneously crystallised out from a mixture of the other pentose aminooxazolines. Aminooxazolines can react with cyanoacetylene in a mild and highly efficient manner, controlled by inorganic phosphate, to give the cytidine ribonucleotides. Photoanomerization with UV light allows for inversion about the 1' anomeric centre to give the correct beta stereochemistry; one problem with this chemistry is the selective phosphorylation of alpha-cytidine at the 2' position. However, in 2009, they showed that the same simple building blocks allow access, via phosphate controlled nucleobase elaboration, to 2',3'-cyclic pyrimidine nucleotides directly, which are known to be able to polymerise into RNA. Organic chemist Donna Blackmond described this finding as "strong evidence" in favour of the RNA world. However, John Sutherland said that while his team's work suggests that nucleic acids played an early and central role in the origin of life, it did not necessarily support the RNA world hypothesis in the strict sense, which he described as a "restrictive, hypothetical arrangement".

The Sutherland group's 2009 paper also highlighted the possibility for the photo-sanitization of the pyrimidine-2',3'-cyclic phosphates. A potential weakness of these routes is the generation of enantioenriched glyceraldehyde, or its 3-phosphate derivative (glyceraldehyde prefers to exist as its keto tautomer dihydroxyacetone).Wikipedia:Citation needed

On August 8, 2011, a report, based on NASA studies with meteorites found on Earth, was published suggesting building blocks of RNA (adenine, guanine and related organic molecules) may have been formed extraterrestrially in outer space. In 2017, a numerical model suggests that the RNA world may have emerged in warm ponds on the early Earth, and that meteorites were a plausible and probable source of the RNA building blocks (ribose and nucleic acids) to these environments. On August 29, 2012, astronomers at Copenhagen University reported the detection of a specific sugar molecule, glycolaldehyde, in a distant star system. The molecule was found around the protostellar binary *IRAS 16293-2422*, which is located 400 light years from Earth. Because glycolaldehyde is needed to form RNA, this finding suggests that complex organic molecules may form in stellar systems prior to the formation of planets, eventually arriving on young planets early in their formation.

Prebiotic RNA synthesis

Nucleotides are the fundamental molecules that combine in series to form RNA. They consist of a nitrogenous base attached to a sugar-phosphate backbone. RNA is made of long stretches of specific nucleotides arranged so that their sequence of bases carries information. The RNA world hypothesis holds that in the primordial soup (or sandwich), there existed free-floating nucleotides. These nucleotides regularly formed bonds with one another, which often broke because the change in energy was so low. However, certain sequences of base pairs have catalytic properties that lower the energy of their chain being created, enabling them to stay together for longer periods of time. As each chain grew longer, it attracted more matching nucleotides faster, causing chains to now form faster than they were breaking down.

These chains have been proposed by some as the first, primitive forms of life. In an RNA world, different sets of RNA strands would have had different replication outputs, which would have increased or decreased their frequency in the population, i.e. natural selection. As the fittest sets of RNA molecules expanded their numbers, novel catalytic properties added by mutation, which benefitted their persistence and expansion, could accumulate in the population. Such an autocatalytic set of ribozymes, capable of self replication in about an hour, has been identified. It was produced by molecular competition (*in vitro* evolution) of candidate enzyme mixtures.

Competition between RNA may have favored the emergence of cooperation between different RNA chains, opening the way for the formation of the first protocell. Eventually, RNA chains developed with catalytic properties that help amino acids bind together (a process called peptide-bonding). These amino acids could then assist with RNA synthesis, giving those RNA chains that could serve as ribozymes the selective advantage. The ability to catalyze one step in protein synthesis, aminoacylation of RNA, has been demonstrated in a short (five-nucleotide) segment of RNA.

In March 2015, NASA scientists reported that, for the first time, complex DNA and RNA organic compounds of life, including uracil, cytosine and thymine, have been formed in the laboratory under conditions found only in outer space, using starting chemicals, like pyrimidine, found in meteorites. Pyrimidine, like polycyclic aromatic hydrocarbons (PAHs), may have been formed in giant red stars or in interstellar dust and gas clouds, according to the scientists.

Evolution of DNA

One of the problems with the RNA world hypothesis is to discover the pathway by which RNA became upgraded to the DNA system. Geoffrey Diemer and Ken Stedman, at Portland State University in Oregon, may have found a solution. While conducting a survey of viruses in a hot acidic lake in Lassen Volcanic National Park, California, they uncovered evidence that a simple DNA virus had acquired a gene from a completely unrelated RNA-based virus. Virologist Luis Villareal of the University of California Irvine also suggests that viruses capable of converting an RNA-based gene into DNA and then incorporating it into a more complex DNA-based genome might have been common in the Virus world during the RNA to DNA transition some 4 billion years ago.[251] This finding bolsters the argument for the transfer of information from the RNA world to the emerging DNA world before the emergence of the last universal common ancestor. From the research, the diversity of this virus world is still with us.

Viroids

Additional evidence supporting the concept of an RNA world has resulted from research on viroids, the first representatives of a novel domain of "subviral pathogens". Viroids are mostly plant pathogens, which consist of short stretches (a few hundred nucleobases) of highly complementary, circular, single-stranded, and non-coding RNA without a protein coat. Compared with other infectious plant pathogens, viroids are extremely small in size, ranging from 246 to 467 nucleobases. In comparison, the genome of the smallest known viruses capable of causing an infection are about 2,000 nucleobases long.

In 1989, Diener proposed that, based on their characteristic properties, viroids are more plausible "living relics" of the RNA world than are introns or other RNAs then so considered. If so, viroids have attained potential significance beyond plant pathology to evolutionary biology, by representing the most plausible macromolecules known capable of explaining crucial intermediate steps in the evolution of life from inanimate matter (see: abiogenesis).

Apparently, Diener's hypothesis lay dormant until 2014, when Flores et al. published a review paper, in which Diener's evidence supporting his hypothesis was summarized. In the same year, a New York Times science writer published a popularized version of Diener's proposal, in which, however, he mistakenly credited Flores et al. with the hypothesis' original conception.

Pertinent viroid properties listed in 1989 are:

1. their small size, imposed by error-prone replication;
2. their high guanine and cytosine content, which increases stability and replication fidelity;
3. their circular structure, which assures complete replication without genomic tags;
4. existence of structural periodicity, which permits modular assembly into enlarged genomes;
5. their lack of protein-coding ability, consistent with a ribosome-free habitat; and
6. replication mediated in some by ribozymes—the fingerprint of the RNA world.

The existence, in extant cells, of RNAs with molecular properties predicted for RNAs of the RNA World constitutes an additional argument supporting the RNA World hypothesis.

Origin of sex

Eigen *et al.* and Woese[252] proposed that the genomes of early protocells were composed of single-stranded RNA, and that individual genes corresponded to separate RNA segments, rather than being linked end-to-end as in present-day DNA genomes. A protocell that was haploid (one copy of each RNA gene) would be vulnerable to damage, since a single lesion in any RNA segment would be potentially lethal to the protocell (e.g. by blocking replication or inhibiting the function of an essential gene).

Vulnerability to damage could be reduced by maintaining two or more copies of each RNA segment in each protocell, i.e. by maintaining diploidy or polyploidy. Genome redundancy would allow a damaged RNA segment to be replaced by an additional replication of its homolog. However, for such a simple organism, the proportion of available resources tied up in the genetic material would be a large fraction of the total resource budget. Under limited resource conditions, the protocell reproductive rate would likely be inversely related to ploidy number. The protocell's fitness would be reduced by the costs of redundancy. Consequently, coping with damaged RNA genes while minimizing the costs of redundancy would likely have been a fundamental problem for early protocells.

A cost-benefit analysis was carried out in which the costs of maintaining redundancy were balanced against the costs of genome damage. This analysis led to the conclusion that, under a wide range of circumstances, the selected strategy would be for each protocell to be haploid, but to periodically fuse with another haploid protocell to form a transient diploid. The retention of the haploid state maximizes the growth rate. The periodic fusions permit mutual reactivation of otherwise lethally damaged protocells. If at least one damage-free copy of each RNA gene is present in the transient diploid, viable progeny can be formed. For two, rather than one, viable daughter cells to be produced would require an extra replication of the intact RNA gene homologous to any RNA gene that had been damaged prior to the division of the fused protocell. The cycle of haploid reproduction, with occasional fusion to a transient diploid state, followed by splitting to the haploid state, can be considered to be the sexual cycle in its most primitive form.[253] In the absence of this sexual cycle, haploid protocells with damage in an essential RNA gene would simply die.

This model for the early sexual cycle is hypothetical, but it is very similar to the known sexual behavior of the segmented RNA viruses, which are among the simplest organisms known. Influenza virus, whose genome consists of 8 physically separated single-stranded RNA segments, is an example of this type of virus. In segmented RNA viruses, "mating" can occur when a host cell is infected by at least two virus particles. If these viruses each contain an RNA segment with a lethal damage, multiple infection can lead to reactivation providing that at least one undamaged copy of each virus gene is present in the infected cell. This phenomenon is known as "multiplicity reactivation". Multiplicity reactivation has been reported to occur in influenza virus infections after induction of RNA damage by UV-irradiation, and ionizing radiation.

Further developments

Patrick Forterre has been working on a novel hypothesis, called "three viruses, three domains": that viruses were instrumental in the transition from RNA to DNA and the evolution of Bacteria, Archaea, and Eukaryota. He believes the last universal common ancestor was RNA-based and evolved RNA viruses. Some of the viruses evolved into DNA viruses to protect their genes from attack. Through the process of viral infection into hosts the three domains of life evolved. Another interesting proposal is the idea that RNA synthesis might have been driven by temperature gradients, in the process of thermosynthesis. Single nucleotides have been shown to catalyze organic reactions.

Steven Benner has argued that chemical conditions on the planet Mars, such as the presence of boron, molybdenum and oxygen, may have been better for initially producing RNA molecules than those on Earth. If so, life-suitable molecules, originating on Mars, may have later migrated to Earth via panspermia or similar process.

Alternative hypotheses

The hypothesized existence of an RNA world does not exclude a "Pre-RNA world", where a metabolic system based on a different nucleic acid is proposed to pre-date RNA. A candidate nucleic acid is peptide nucleic acid (PNA), which uses simple peptide bonds to link nucleobases. PNA is more stable than RNA, but its ability to be generated under prebiological conditions has yet to be demonstrated experimentally.

Threose nucleic acid (TNA) has also been proposed as a starting point, as has glycol nucleic acid (GNA), and like PNA, also lack experimental evidence for their respective abiogenesis.

An alternative — or complementary — theory of RNA origin is proposed in the PAH world hypothesis, whereby polycyclic aromatic hydrocarbons (PAHs) mediate the synthesis of RNA molecules.[254] PAHs are the most common and abundant of the known polyatomic molecules in the visible Universe, and are a likely constituent of the primordial sea.[255] PAHs and fullerenes (also implicated in the origin of life) have been detected in nebulae.

The iron-sulfur world theory proposes that simple metabolic processes developed before genetic materials did, and these energy-producing cycles catalyzed the production of genes.

Some of the difficulties of producing the precursors on earth are bypassed by another alternative or complementary theory for their origin, panspermia. It discusses the possibility that the earliest life on this planet was carried here from somewhere else in the galaxy, possibly on meteorites similar to the Murchison meteorite. This does not invalidate the concept of an RNA world, but posits that this world or its precursors originated not on Earth but rather another, probably older, planet.

There are hypotheses that are in direct conflict to the RNA world hypothesis. The relative chemical complexity of the nucleotide and the unlikelihood of it spontaneously arising, along with the limited number of combinations possible among four base forms, as well as the need for RNA polymers of some length before seeing enzymatic activity, have led some to reject the RNA world hypothesis in favor of a metabolism-first hypothesis, where the chemistry underlying cellular function arose first, along with the ability to replicate and facilitate this metabolism.

RNA-peptide coevolution

Another proposal is that the dual-molecule system we see today, where a nucleotide-based molecule is needed to synthesize protein, and a peptide-based (protein) molecule is needed to make nucleic acid polymers, represents the original form of life. This theory is called RNA-peptide coevolution, or the Peptide-RNA world, and offers a possible explanation for the rapid evolution of high-quality replication in RNA (since proteins are catalysts), with the disadvantage of having to postulate the coincident formation of two complex molecules, an enzyme (from peptides) and a RNA (from nucleotides). In this Peptide-RNA World scenario, RNA would have contained the instructions for life, while peptides (simple protein enzymes) would have accelerated key chemical reactions to carry out those instructions. The study leaves open the question of exactly how those primitive systems managed to replicate themselves — something neither the RNA World hypothesis nor the Peptide-RNA World theory can yet explain, unless polymerases (enzymes that rapidly assemble the RNA molecule) played a role.

A research project completed in March 2015 by the Sutherland group found that a network of reactions beginning with hydrogen cyanide and hydrogen sulfide, in streams of water irradiated by UV light, could produce the chemical components of proteins and lipids, alongside those of RNA. The researchers used the term "cyanosulfidic" to describe this network of reactions. In November 2017, a team at the Scripps Research Institute identified reactions involving the compound diamidophosphate which could have linked the chemical components into short peptide and lipid chains as well as short RNA-like chains of nucleotides.[256]

Implications of the RNA world

The RNA world hypothesis, if true, has important implications for the definition of life. For most of the time that followed Watson and Crick's elucidation of DNA structure in 1953, life was largely defined in terms of DNA and proteins: DNA and proteins seemed the dominant macromolecules in the living cell, with RNA only aiding in creating proteins from the DNA blueprint.

The RNA world hypothesis places RNA at center-stage when life originated. The RNA world hypothesis is supported by the observations that ribosomes are ribozymes: the catalytic site is composed of RNA, and proteins hold no major structural role and are of peripheral functional importance. This was confirmed with the deciphering of the 3-dimensional structure of the ribosome in 2001. Specifically, peptide bond formation, the reaction that binds amino acids together into proteins, is now known to be catalyzed by an adenine residue in the rRNA.

RNAs are known to play roles in other cellular catalytic processes, specifically in the targeting of enzymes to specific RNA sequences. In eukaryotes, the processing of pre-mRNA and RNA editing take place at sites determined by the base pairing between the target RNA and RNA constituents of small nuclear ribonucleoproteins (snRNPs). Such enzyme targeting is also responsible for gene down regulation though RNA interference (RNAi), where an enzyme-associated guide RNA targets specific mRNA for selective destruction. Likewise, in eukaryotes the maintenance of telomeres involves copying of an RNA template that is a constituent part of the telomerase ribonucleoprotein enzyme. Another cellular organelle, the vault, includes a ribonucleoprotein component, although the function of this organelle remains to be elucidated.

Further reading

<templatestyles src="Template:Refbegin/styles.css" />

- Attwater J, Raguram A, Morgunov AS, Gianni E, Holliger P (15 May 2018). "Ribozyme-catalysed RNA synthesis using triplet building blocks". *eLife*. 7: e35255. doi: 10.7554/eLife.35255[257]. PMID 29759114[258].
- Cairns-Smith AG (1993). *Genetic Takeover: And the Mineral Origins of Life*. Cambridge University Press. ISBN 0-521-23312-7.

- Orgel LE (Oct 1994). "The origin of life on the earth". *Scientific American*. **271** (4): 76–83. Bibcode: 1994SciAm.271d..76O[259]. doi: 10.1038/scientificamerican1094-76[260]. PMID 7524147[261].
- Orgel LE (2004). "Prebiotic chemistry and the origin of the RNA world". *Critical Reviews in Biochemistry and Molecular Biology*. **39** (2): 99–123. doi: 10.1080/10409230490460765[262]. PMID 15217990[263].
- Woolfson A (September 2000). *Life Without Genes*. London: Flamingo. ISBN 978-0-00-654874-4.
- Vlassov AV, Kazakov SA, Johnston BH, Landweber LF (Aug 2005). "The RNA world on ice: a new scenario for the emergence of RNA information". *Journal of Molecular Evolution*. **61** (2): 264–73. Bibcode: 2005JMolE..61..264V[264]. doi: 10.1007/s00239-004-0362-7[265]. PMID 16044244[266].
- Engelhart AE, Hud NV (Dec 2010). "Primitive genetic polymers"[267] (PDF). *Cold Spring Harbor Perspectives in Biology*. **2** (12): a002196. doi: 10.1101/cshperspect.a002196[268]. PMC 2982173[269] ∂. PMID 20462999[270].
- Bernhardt HS (2012). "The RNA world hypothesis: the worst theory of the early evolution of life (except for all the others)(a)"[271]. *Biology Direct*. **7** (1): 23. doi: 10.1186/1745-6150-7-23[272]. PMC 3495036[273] ∂. PMID 22793875[274].
- Sutherland JD (Apr 2010). "Ribonucleotides"[275]. *Cold Spring Harbor Perspectives in Biology*. **2** (4): a005439. doi: 10.1101/cshperspect.a005439[276]. PMC 2845210[275] ∂. PMID 20452951[277].

External links

- "Understanding the RNA World"[278]. *Exploring Life's Origins*. Exploring Origins Project.
- Ferris, James P. "The Formation of the RNA World"[279]. The New York Center for Studies of the Origins of Life, Rensselaer Polytechnic Institute. Archived from the original[280] on 2012-03-01.
- Altman, Sidney (2001). "The RNA World"[281]. *NobelPrize.org*. Nobel Media.
- Kuska, Robert (June 2002). "A World Apart"[282] (PDF). *HHMI Bulletin*. Howard Hughes Medical Institute. pp. 14–19. Archived[283] (PDF) from the original on 2004-05-22.
- Cech, Thomas R. (2004). "Exploring the New RNA World"[284]. *NobelPrize.org*. Nobel Media.
- Sutherland, J. D. (April 2010). "Ribonucleotides"[275]. *Cold Spring Harbor Perspectives in Biology*. **2** (4): a005439. doi: 10.1101/cshperspect.a005439[276]. PMC 2845210[275] ∂. PMID 20452951[277].

Origin of biological metabolism

Iron–sulfur world hypothesis

The **iron–sulfur world hypothesis** is a set of proposals for the origin of life and the early evolution of life advanced in a series of articles between 1988 and 1992 by Günter Wächtershäuser, a Munich patent lawyer with a degree in chemistry, who had been encouraged and supported by philosopher Karl R. Popper to publish his ideas. The hypothesis proposes that early life may have formed on the surface of iron sulfide minerals, hence the name. It was developed by retrodiction from extant biochemistry in conjunction with chemical experiments.

Origin of life

Pioneer organism

Wächtershäuser proposes that the earliest form of life, termed "pioneer organism", originated in a volcanic hydrothermal flow at high pressure and high (100 °C) temperature. It had a composite structure of a mineral base with catalytic transition metal centers (predominantly iron and nickel, but also perhaps cobalt, manganese, tungsten and zinc). The catalytic centers catalyzed autotrophic carbon fixation pathways generating small molecule (non-polymer) organic compounds from inorganic gases (e.g. carbon monoxide, carbon dioxide, hydrogen cyanide and hydrogen sulfide). These organic compounds were retained on or in the mineral base as organic ligands of the transition metal centers with a flow retention time in correspondence with their mineral bonding strength thereby defining an autocatalytic "surface metabolism". The catalytic transition metal centers became autocatalytic by being accelerated by their organic products turned ligands. The carbon fixation metabolism became autocatalytic by forming a metabolic cycle in the form of a primitive sulfur-dependent version of the reductive citric acid cycle. Accelerated catalysts expanded the metabolism and new metabolic products further accelerated the catalysts. The idea is that once such a primitive autocatalytic metabolism was established, its intrinsically synthetic chemistry began to produce ever more complex organic compounds, ever more complex pathways and ever more complex catalytic centers.

Nutrient conversions

The water gas shift reaction ($CO + H_2O \rightarrow CO_2 + H_2$) occurs in volcanic fluids with diverse catalysts or without catalysts. The combination of ferrous sulfide and hydrogen sulfide as reducing agents in conjunction with pyrite formation – $FeS + H_2S \rightarrow FeS_2 + 2H^+ + 2e^-$ (or H_2 instead of $2H^+ + 2e^-$) – has been demonstrated under mild volcanic conditions. This key result has been disputed. Nitrogen fixation has been demonstrated for the isotope $^{15}N_2$ in conjunction with pyrite formation. Ammonia forms from nitrate with FeS/H_2S as reductant. Methylmercaptan [CH_3-SH] and carbon oxysulfide [COS] form from CO_2 and FeS/H_2S, or from CO and H_2 in the presence of NiS.

Synthetic reactions

Reaction of carbon monoxide (CO), hydrogen sulfide (H_2S) and methanethiol CH_3SH in the presence of nickel sulfide and iron sulfide generates the methyl thioester of acetic acid [CH_3-CO-SCH$_3$] and presumably thioacetic acid (CH_3-CO-SH) as the simplest activated acetic acid analogues of acetyl-CoA. These activated acetic acid derivatives serve as starting materials for subsequent exergonic synthetic steps. They also serve for energy coupling with endergonic reactions, notably the formation of (phospho)anhydride compounds. However, Huber and Wächtershäuser reported a low 0.5% acetate yields based on input of CH_3SH (Methanethiol) (8 mM) in the presence of 350 mM CO. This is about 500 times and 3700 times the highest CH_3SH and CO concentrations respectively measured to date in a natural hydrothermal vent fluid.

Reaction of nickel hydroxide with hydrogen cyanide (HCN) (in the presence or absence of ferrous hydroxide, hydrogen sulfide or methyl mercaptan) generates nickel cyanide, which reacts with carbon monoxide (CO) to generate pairs of α-hydroxy and α-amino acids: e.g. glycolate/glycine, lactate/alanine, glycerate/serine; as well as pyruvic acid in significant quantities. Pyruvic acid is also formed at high pressure and high temperature from CO, H_2O, FeS in the presence of nonyl mercaptan. Reaction of pyruvic acid or other α-keto acids with ammonia in the presence of ferrous hydroxide or in the presence of ferrous sulfide and hydrogen sulfide generates alanine or other α-amino acids. Reaction of α-amino acids in aqueous solution with COS or with CO and H_2S

generates a peptide cycle wherein dipeptides, tripeptides etc. are formed and subsequently degraded via N-terminal hydantoin moieties and N-terminal urea moieties and subsequent cleavage of the N-terminal amino acid unit.[285]

Proposed reaction mechanism for reduction of CO_2 on FeS: Ying et al. (2007) have proved that direct transformation of mackinawite (FeS) to pyrite (FeS$_2$) on reaction with H_2S till 300 °C is not possible without the presence of critical amount of oxidant. In the absence of any oxidant, FeS reacts with H_2S up to 300 °C to give pyrrhotite. Farid et al. have proved experimentally that mackinawite (FeS) has ability to reduce CO_2 to CO at temperature higher than 300 °C. They reported that the surface of FeS is oxidized, which on reaction with H_2S gives pyrite (FeS$_2$). It is expected that CO reacts with H_2O in the Drobner experiment to give H_2.

Early evolution

Early evolution is defined as beginning with the origin of life and ending with the last universal common ancestor (LUCA). According to the iron–sulfur world theory it covers a coevolution of cellular organization (cellularization), the genetic machinery and enzymatization of the metabolism.

Cellularization

Cellularization occurs in several stages. It begins with the formation of primitive lipids (e.g. fatty acids or isoprenoid acids) in the surface metabolism. These lipids accumulate on or in the mineral base. This lipophilizes the outer or inner surfaces of the mineral base, which promotes condensation reactions over hydrolytic reactions by lowering the activity of water and protons.

In the next stage lipid membranes are formed. While still anchored to the mineral base they form a semi-cell bounded partly by the mineral base and partly by the membrane. Further lipid evolution leads to self-supporting lipid membranes and closed cells. The earliest closed cells are pre-cells (*sensu* Kandler) because they allow frequent exchange of genetic material (e.g. by fusions). According to Woese, this frequent exchange of genetic material is the cause for the existence of the common stem in the tree of life and for a very rapid early evolution.

Proto-ecological systems

William Martin and Michael Russell suggest that the first cellular life forms may have evolved inside alkaline hydrothermal vents at seafloor spreading zones in the deep sea. These structures consist of microscale caverns that are coated by thin membraneous metal sulfide walls. Therefore, these structures would resolve several critical points germane to Wächtershäuser's suggestions at once:

1. the micro-caverns provide a means of concentrating newly synthesised molecules, thereby increasing the chance of forming oligomers;
2. the steep temperature gradients inside the hydrothermal vent allow for establishing "optimum zones" of partial reactions in different regions of the vent (e.g. monomer synthesis in the hotter, oligomerisation in the colder parts);
3. the flow of hydrothermal water through the structure provides a constant source of building blocks and energy (chemical disequilibrium between hydrothermal hydrogen and marine carbon dioxide);
4. the model allows for a succession of different steps of cellular evolution (prebiotic chemistry, monomer and oligomer synthesis, peptide and protein synthesis, RNA world, ribonucleoprotein assembly and DNA world) in a single structure, facilitating exchange between all developmental stages;
5. synthesis of lipids as a means of "closing" the cells against the environment is not necessary, until basically all cellular functions are developed.

This model locates the "last universal common ancestor" (LUCA) within the inorganically formed physical confines of an alkaline hydrothermal vent, rather than assuming the existence of a free-living form of LUCA. The last evolutionary step en route to bona fide free-living cells would be the synthesis of a lipid membrane that finally allows the organisms to leave the microcavern system of the vent. This postulated late acquisition of the biosynthesis of lipids as directed by genetically encoded peptides is consistent with the presence of completely different types of membrane lipids in archaea and bacteria (plus eukaryotes). The kind of vent at the foreground of their suggestion is chemically more similar to the warm (ca. 100 °C) off ridge vents such as Lost City than to the more familiar black smoker type vents (ca. 350 °C).

In an abiotic world, a thermocline of temperatures and a chemocline in concentration is associated with the pre-biotic synthesis of organic molecules, hotter in proximity to the chemically rich vent, cooler but also less chemically rich at greater distances. The migration of synthesized compounds from areas of high concentration to areas of low concentration gives a directionality that provides both source and sink in a self-organizing fashion, enabling a proto-metabolic process by which acetic acid production and its eventual oxidization can be spatially organized.

In this way many of the individual reactions that are today found in central metabolism could initially have occurred independent of any developing cell membrane. Each vent microcompartment is functionally equivalent to a single cell. Chemical communities having greater structural integrity and resilience to wildly fluctuating conditions are then selected for; their success would lead to local zones of depletion for important precursor chemicals. Progressive incorporation of these precursor components within a cell membrane would gradually increase metabolic complexity within the cell membrane, whilst leading to greater environmental simplicity in the external environment. In principle, this could lead to the development of complex catalytic sets capable of self-maintenance.

Russell adds a significant factor to these ideas, by pointing out that semipermeable mackinawite (an iron sulfide mineral) and silicate membranes could naturally develop under these conditions and electrochemically link reactions separated in space, if not in time.[286]

Panspermia

Panspermia (from Ancient Greek πᾶν *(pan)*, meaning 'all', and σπέρμα *(sperma)*, meaning 'seed') is the hypothesis that life exists throughout the Universe, distributed by space dust, meteoroids, asteroids, comets, planetoids, and also by spacecraft carrying unintended contamination by microorganisms.[287]

Panspermia hypotheses propose (for example) that microscopic life-forms that can survive the effects of space (such as extremophiles) can become trapped in debris ejected into space after collisions between planets and small Solar System bodies that harbor life. Some organisms may travel dormant for an extended amount of time before colliding randomly with other planets or intermingling with protoplanetary disks. Under certain ideal impact circumstances (into a body of water, for example), and ideal conditions on a new planet's surfaces, it is possible that the surviving organisms could become active and begin to colonize their new environment. Panspermia studies concentrate not on how life began, but on the methods that may cause its distribution in the Universe.[288,289,290]

Pseudo-panspermia (sometimes called *"soft panspermia"* or *"molecular panspermia"*) argues that the pre-biotic organic building-blocks of life originated in space, became incorporated in the solar nebula from which planets condensed, and were further—and continuously—distributed to planetary surfaces where life then emerged (abiogenesis). From the early 1970s it started

Figure 56: *Panspermia proposes that bodies such as comets transported life forms such as bacteria - complete with their DNA - through space to the Earth.*

to become evident that interstellar dust included a large component of organic molecules. Interstellar molecules are formed by chemical reactions within very sparse interstellar or circumstellar clouds of dust and gas. The dust plays a critical role in shielding the molecules from the ionizing effect of ultraviolet radiation emitted by stars.

The chemistry leading to life may have begun shortly after the Big Bang, 13.8 billion years ago, during a habitable epoch when the Universe was only 10 to 17 million years old. Though the presence of life is confirmed only on the Earth, some scientists think that extraterrestrial life is not only plausible, but probable or inevitable. Probes and instruments have started examining other planets and moons in the Solar System and in other planetary systems for evidence of having once supported simple life, and projects such as SETI attempt to detect radio transmissions from possible extra-terrestrial civilizations.

History

The first known mention of the term was in the writings of the 5th-century BC Greek philosopher Anaxagoras.[291] Panspermia began to assume a more scientific form through the proposals of Jöns Jacob Berzelius (1834), Hermann E. Richter (1865), Kelvin (1871), Hermann von Helmholtz (1879) and finally reaching the level of a detailed scientific hypothesis through the efforts of the Swedish chemist Svante Arrhenius (1903).[292]

Fred Hoyle (1915–2001) and Chandra Wickramasinghe (born 1939) were influential proponents of panspermia. In 1974 they proposed the hypothesis that some dust in interstellar space was largely organic (containing carbon), which Wickramasinghe later proved to be correct. Hoyle and Wickramasinghe further contended that life forms continue to enter the Earth's atmosphere, and may be responsible for epidemic outbreaks, new diseases, and the genetic novelty necessary for macroevolution.

In an Origins Symposium presentation on April 7, 2009, physicist Stephen Hawking stated his opinion about what humans may find when venturing into space, such as the possibility of alien life through the theory of panspermia: "Life could spread from planet to planet or from stellar system to stellar system, carried on meteors."

Three series of astrobiology experiments have been conducted outside the International Space Station between 2008 and 2015 (EXPOSE) where a wide variety of biomolecules, microorganisms, and their spores were exposed to the solar flux and vacuum of space for about 1.5 years. Some organisms survived in an inactive state for considerable lengths of time, and those samples sheltered by simulated meteorite material provide experimental evidence for the likelihood of the hypothetical scenario of lithopanspermia.

Several simulations in laboratories and in low Earth orbit suggest that ejection, entry and impact is survivable for some simple organisms. In 2015, remains of biotic material were found in 4.1 billion-year-old rocks in Western Australia, when the young Earth was about 400 million years old.[293] According to one researcher, "If life arose relatively quickly on Earth ... then it could be common in the universe."

In April 2018 a Russian team published a paper which disclosed that they found DNA on the exterior of the ISS from land and marine bacteria similar to those previously observed in superficial micro layers at the Barents and Kara seas' coastal zones. They conclude "The presence of the wild land and marine bacteria DNA on the ISS suggests their possible transfer from the stratosphere into the ionosphere with the ascending branch of the global atmospheric electrical circuit. Alternatively, the wild land and marine bacteria as well as the ISS bacteria may all have an ultimate space origin."

Proposed mechanisms

Panspermia can be said to be either interstellar (between star systems) or interplanetary (between planets in the same star system); its transport mechanisms may include comets, radiation pressure and lithopanspermia (microorganisms embedded in rocks). Interplanetary transfer of nonliving material is well documented, as evidenced by meteorites of Martian origin found on Earth. Space probes may also be a viable transport mechanism for interplanetary cross-pollination in the Solar System or even beyond. However, space agencies have implemented planetary protection procedures to reduce the risk of planetary contamination,[294,295] although, as recently discovered, some microorganisms, such as Tersicoccus phoenicis, may be resistant to procedures used in spacecraft assembly clean room facilities. In 2012, mathematician Edward Belbruno and astronomers Amaya Moro-Martín and Renu Malhotra proposed that gravitational low-energy transfer of rocks among the young planets of stars in their birth cluster is commonplace, and not rare in the general galactic stellar population.[296] Deliberate directed panspermia from space to seed Earth or sent from Earth to seed other planetary systems have also been proposed. One twist to the hypothesis by engineer Thomas Dehel (2006), proposes that plasmoid magnetic fields ejected from the magnetosphere may move the few spores lifted from the Earth's atmosphere with sufficient speed to cross interstellar space to other systems before the spores can be destroyed.

Radiopanspermia

In 1903, Svante Arrhenius published in his article *The Distribution of Life in Space*,[297] the hypothesis now called radiopanspermia, that microscopic forms of life can be propagated in space, driven by the radiation pressure from stars. Arrhenius argued that particles at a critical size below 1.5 μm would be propagated at high speed by radiation pressure of the Sun. However, because its effectiveness decreases with increasing size of the particle, this mechanism holds for very tiny particles only, such as single bacterial spores. The main criticism of radiopanspermia hypothesis came from Iosif Shklovsky and Carl Sagan, who pointed out the proofs of the lethal action of space radiations (UV and X-rays) in the cosmos. Regardless of the evidence, Wallis and Wickramasinghe argued in 2004 that the transport of individual bacteria or clumps of bacteria, is overwhelmingly more important than lithopanspermia in terms of numbers of microbes transferred, even accounting for the death rate of unprotected bacteria in transit.

Then, data gathered by the orbital experiments ERA, BIOPAN, EXOSTACK and EXPOSE, determined that isolated spores, including those of *B. subtilis*, were killed by several orders of magnitude if exposed to the full space environment for a mere few seconds, but if shielded against solar UV, the spores were capable of surviving in space for up to six years while embedded in clay or meteorite powder (artificial meteorites). Though minimal protection is required to shelter a spore against UV radiation, exposure to solar UV and cosmic ionizing radiation of unprotected DNA, break it up into its bases. Also, exposing

DNA to the ultrahigh vacuum of space alone is sufficient to cause DNA damage, so the transport of unprotected DNA or RNA during interplanetary flights powered solely by light pressure is extremely unlikely. The feasibility of other means of transport for the more massive shielded spores into the outer Solar System – for example, through gravitational capture by comets – is at this time unknown.

Based on experimental data on radiation effects and DNA stability, it has been concluded that for such long travel times, boulder-sized rocks which are greater than or equal to 1 meter in diameter are required to effectively shield resistant microorganisms, such as bacterial spores against galactic cosmic radiation. These results clearly negate the radiopanspermia hypothesis, which requires single spores accelerated by the radiation pressure of the Sun, requiring many years to travel between the planets, and support the likelihood of interplanetary transfer of microorganisms within asteroids or comets, the so-called **lithopanspermia** hypothesis.

Lithopanspermia

Lithopanspermia, the transfer of organisms in rocks from one planet to another either through interplanetary or interstellar space, remains speculative. Although there is no evidence that lithopanspermia has occurred in the Solar System, the various stages have become amenable to experimental testing.

- **Planetary ejection** — For lithopanspermia to occur, researchers have suggested that microorganisms must survive ejection from a planetary surface which involves extreme forces of acceleration and shock with associated temperature excursions. Hypothetical values of shock pressures experienced by ejected rocks are obtained with Martian meteorites, which suggest the shock pressures of approximately 5 to 55 GPa, acceleration of 3 Mm/s^2 and jerk of 6 Gm/s^3 and post-shock temperature increases of about 1 K to 1000 K. To determine the effect of acceleration during ejection on microorganisms, rifle and ultracentrifuge methods were successfully used under simulated outer space conditions.
- **Survival in transit** — The survival of microorganisms has been studied extensively using both simulated facilities and in low Earth orbit. A large number of microorganisms have been selected for exposure experiments. It is possible to separate these microorganisms into two groups, the human-borne, and the extremophiles. Studying the human-borne microorganisms is significant for human welfare and future manned missions; whilst the extremophiles are vital for studying the physiological requirements of survival in space.

- **Atmospheric entry** — An important aspect of the lithopanspermia hypothesis to test is that microbes situated on or within rocks could survive hypervelocity entry from space through Earth's atmosphere (Cockell, 2008). As with planetary ejection, this is experimentally tractable, with sounding rockets and orbital vehicles being used for microbiological experiments. *B. subtilis* spores inoculated onto granite domes were subjected to hypervelocity atmospheric transit (twice) by launch to a ∼120 km altitude on an Orion two-stage rocket. The spores were shown to have survived on the sides of the rock, but they did not survive on the forward-facing surface that was subjected to a maximum temperature of 145 °C. In separate experiments, as part of the ESA STONE experiment, numerous organisms were embedded in different types or rocks and were mounted in the heat shield of six Foton re-entry capsules. During reentry, the rock samples were subjected to temperatures and pressure loads comparable to those experienced in meteorites. The exogenous arrival of photosynthetic microorganisms could have quite profound consequences for the course of biological evolution on the inoculated planet. As photosynthetic organisms must be close to the surface of a rock to obtain sufficient light energy, atmospheric transit might act as a filter against them by ablating the surface layers of the rock. Although cyanobacteria have been shown to survive the desiccating, freezing conditions of space in orbital experiments, this would be of no benefit as the STONE experiment showed that they cannot survive atmospheric entry. Thus, non-photosynthetic organisms deep within rocks have a chance to survive the exit and entry process. (See also: Impact survival.) Research presented at the European Planetary Science Congress in 2015 suggests that ejection, entry and impact is survivable for some simple organisms.

Accidental panspermia

Thomas Gold, a professor of astronomy, suggested in 1960 the hypothesis of "Cosmic Garbage", that life on Earth might have originated accidentally from a pile of waste products dumped on Earth long ago by extraterrestrial beings.[298]

Directed panspermia

Directed panspermia concerns the deliberate transport of microorganisms in space, sent to Earth to start life here, or sent from Earth to seed new planetary systems with life by introduced species of microorganisms on lifeless planets. The Nobel prize winner Francis Crick, along with Leslie Orgel proposed that life may have been purposely spread by an advanced extraterrestrial civilization, but considering an early "RNA world" Crick noted later that life may have originated on Earth.[299] It has been suggested that 'directed' panspermia was proposed in order to counteract various objections, including the argument that microbes would be inactivated by the space environment and cosmic radiation before they could make a chance encounter with Earth.

Conversely, active directed panspermia has been proposed to secure and expand life in space. This may be motivated by biotic ethics that values, and seeks to propagate, the basic patterns of our organic gene/protein life-form. The panbiotic program would seed new planetary systems nearby, and clusters of new stars in interstellar clouds. These young targets, where local life would not have formed yet, avoid any interference with local life.

For example, microbial payloads launched by solar sails at speeds up to 0.0001 c (30,000 m/s) would reach targets at 10 to 100 light-years in 0.1 million to 1 million years. Fleets of microbial capsules can be aimed at clusters of new stars in star-forming clouds, where they may land on planets or captured by asteroids and comets and later delivered to planets. Payloads may contain extremophiles for diverse environments and cyanobacteria similar to early microorganisms. Hardy multicellular organisms (rotifer cysts) may be included to induce higher evolution.

The probability of hitting the target zone can be calculated from $P(target) = \frac{A(target)}{\pi (dy)^2} = \frac{ar(target)^2 v^2}{(tp)^2 d^4}$ where A(target) is the cross-section of the target area, dy is the positional uncertainty at arrival; a – constant (depending on units), r(target) is the radius of the target area; v the velocity of the probe; (tp) the targeting precision (arcsec/yr); and d the distance to the target, guided by high-resolution astrometry of 1×10^{-5} arcsec/yr (all units in SIU). These calculations show that relatively near target stars(Alpha PsA, Beta Pictoris) can be seeded by milligrams of launched microbes; while seeding the Rho Ophiochus star-forming cloud requires hundreds of kilograms of dispersed capsules.

Directed panspermia to secure and expand life in space is becoming possible because of developments in solar sails, precise astrometry, extrasolar planets, extremophiles and microbial genetic engineering. After determining the composition of chosen meteorites, astroecologists performed laboratory experiments that suggest that many colonizing microorganisms and some plants could obtain many of their chemical nutrients from asteroid and cometary materials. However, the scientists noted that phosphate (PO_4) and nitrate (NO_3–N) critically limit nutrition to many terrestrial lifeforms. With such materials, and energy from long-lived stars, microscopic life planted by directed panspermia could find an immense future in the galaxy.

A number of publications since 1979 have proposed the idea that directed panspermia could be demonstrated to be the origin of all life on Earth if a distinctive 'signature' message were found, deliberately implanted into either the genome or the genetic code of the first microorganisms by our hypothetical progenitor.

In 2013 a team of physicists claimed that they had found mathematical and semiotic patterns in the genetic code which they think is evidence for such a signature. This claim has been refuted by biologist PZ Myers who said, writing in Pharyngula:

> *Unfortunately, what they've so honestly described is good old honest garbage ... Their methods failed to recognize a well-known functional association in the genetic code; they did not rule out the operation of natural law before rushing to falsely infer design ... We certainly don't need to invoke panspermia. Nothing in the genetic code requires design. and the authors haven't demonstrated otherwise.*

In a later peer-reviewed article, the authors address the operation of natural law in an extensive statistical test, and draw the same conclusion as in the previous article. In special sections they also discuss methodological concerns raised by PZ Myers and some others.

Pseudo-panspermia

Pseudo-panspermia (sometimes called soft panspermia, molecular panspermia or quasi-panspermia) proposes that the organic molecules used for life originated in space and were incorporated in the solar nebula, from which the planets condensed and were further —and continuously— distributed to planetary surfaces where life then emerged (abiogenesis). From the early 1970s it was becoming evident that interstellar dust consisted of a large component of organic molecules. The first suggestion came from Chandra Wickramasinghe, who proposed a polymeric composition based on the molecule formaldehyde (CH_2O). Interstellar molecules are formed by chemical reactions within very sparse interstellar or circumstellar clouds of dust and gas. Usually this occurs when a molecule becomes ionized, often as the result of an interaction with cosmic rays. This positively charged molecule then draws in a nearby reactant by electrostatic attraction of the neutral molecule's electrons. Molecules can also be generated by reactions between neutral atoms and molecules, although this process is generally slower. The dust plays a critical role of shielding the molecules from the ionizing effect of ultraviolet radiation emitted by stars.

A 2008 analysis of $^{12}C/^{13}C$ isotopic ratios of organic compounds found in the Murchison meteorite indicates a non-terrestrial origin for these molecules rather than terrestrial contamination. Biologically relevant molecules identified so far include uracil, an RNA nucleobase, and xanthine. These results demonstrate that many organic compounds which are components of life on

Earth were already present in the early Solar System and may have played a key role in life's origin.

In August 2009, NASA scientists identified one of the fundamental chemical building-blocks of life (the amino acid glycine) in a comet for the first time.

In August 2011, a report, based on NASA studies with meteorites found on Earth, was published suggesting building blocks of DNA (adenine, guanine and related organic molecules) may have been formed extraterrestrially in outer space. In October 2011, scientists reported that cosmic dust contains complex organic matter ("amorphous organic solids with a mixed aromatic-aliphatic structure") that could be created naturally, and rapidly, by stars. One of the scientists suggested that these complex organic compounds may have been related to the development of life on Earth and said that, "If this is the case, life on Earth may have had an easier time getting started as these organics can serve as basic ingredients for life."

In August 2012, and in a world first, astronomers at Copenhagen University reported the detection of a specific sugar molecule, glycolaldehyde, in a distant star system. The molecule was found around the protostellar binary *IRAS 16293-2422*, which is located 400 light years from Earth. Glycolaldehyde is needed to form ribonucleic acid, or RNA, which is similar in function to DNA. This finding suggests that complex organic molecules may form in stellar systems prior to the formation of planets, eventually arriving on young planets early in their formation.

In September 2012, NASA scientists reported that polycyclic aromatic hydrocarbons (PAHs), subjected to interstellar medium (ISM) conditions, are transformed, through hydrogenation, oxygenation and hydroxylation, to more complex organics – "a step along the path toward amino acids and nucleotides, the raw materials of proteins and DNA, respectively". Further, as a result of these transformations, the PAHs lose their spectroscopic signature which could be one of the reasons "for the lack of PAH detection in interstellar ice grains, particularly the outer regions of cold, dense clouds or the upper molecular layers of protoplanetary disks."

In 2013, the Atacama Large Millimeter Array (ALMA Project) confirmed that researchers have discovered an important pair of prebiotic molecules in the icy particles in interstellar space (ISM). The chemicals, found in a giant cloud of gas about 25,000 light-years from Earth in ISM, may be a precursor to a key component of DNA and the other may have a role in the formation of an important amino acid. Researchers found a molecule called cyanomethanimine, which produces adenine, one of the four nucleobases that form the "rungs" in the ladder-like structure of DNA. The other molecule, called ethanamine, is thought to play a role in forming alanine, one of the twenty amino acids in the genetic code. Previously, scientists thought such processes took place in the very tenuous gas between the stars. The new discoveries, however, suggest that the chemical formation sequences for these molecules occurred not in gas, but on the surfaces of ice grains in interstellar space. NASA ALMA scientist Anthony Remijan stated that finding these molecules in an interstellar gas cloud means that important building blocks for DNA and amino acids can 'seed' newly formed planets with the chemical precursors for life.[300]

In March 2013, a simulation experiment indicate that dipeptides (pairs of amino acids) that can be building blocks of proteins, can be created in interstellar dust.

In February 2014, NASA announced a greatly upgraded database[301] for tracking polycyclic aromatic hydrocarbons (PAHs) in the universe. According to scientists, more than 20% of the carbon in the universe may be associated with PAHs, possible starting materials for the formation of life. PAHs seem to have been formed shortly after the Big Bang, are widespread throughout the universe, and are associated with new stars and exoplanets.

In March 2015, NASA scientists reported that, for the first time, complex DNA and RNA organic compounds of life, including uracil, cytosine and thymine, have been formed in the laboratory under outer space conditions, using starting chemicals, such as pyrimidine, found in meteorites. Pyrimidine, like polycyclic aromatic hydrocarbons (PAHs), the most carbon-rich chemical found in the Universe, may have been formed in red giants or in interstellar dust and gas clouds, according to the scientists.

In May 2016, the Rosetta Mission team reported the presence of glycine, methylamine and ethylamine in the coma of 67P/Churyumov-Gerasimenko.[302] This, plus the detection of phosphorus, is consistent with the hypothesis that comets played a crucial role in the emergence of life on Earth.

Extraterrestrial life

The chemistry of life may have begun shortly after the Big Bang, 13.8 billion years ago, during a habitable epoch when the Universe was only 10–17 million years old. According to the panspermia hypothesis, microscopic life—distributed by meteoroids, asteroids and other small Solar System bodies—may exist throughout the universe. Nonetheless, Earth is the only place in the universe known by humans to harbor life. The sheer number of planets in the Milky Way galaxy, however, may make it probable that life has arisen somewhere else in the galaxy and the universe. It is generally agreed that the conditions required for the evolution of intelligent life as we know it are probably exceedingly rare in the universe, while simultaneously noting that simple single-celled microorganisms may be more likely.

The extrasolar planet results from the Kepler mission estimate 100–400 billion exoplanets, with over 3,500 as candidates or confirmed exoplanets. On 4 November 2013, astronomers reported, based on Kepler space mission data, that there could be as many as 40 billion Earth-sized planets orbiting in the habitable zones of sun-like stars and red dwarf stars within the Milky Way Galaxy. 11 billion of these estimated planets may be orbiting sun-like stars. The nearest such planet may be 12 light-years away, according to the scientists.

It is estimated that space travel over cosmic distances would take an incredibly long time to an outside observer, and with vast amounts of energy required. However, there are reasons to hypothesize that faster-than-light interstellar space travel might be feasible. This has been explored by NASA scientists since at least 1995.

Hypotheses on extraterrestrial sources of illnesses

Hoyle and Wickramasinghe have speculated that several outbreaks of illnesses on Earth are of extraterrestrial origins, including the 1918 flu pandemic, and certain outbreaks of polio and mad cow disease. For the 1918 flu pandemic they hypothesized that cometary dust brought the virus to Earth simultaneously at multiple locations—a view almost universally dismissed by experts on this pandemic. Hoyle also speculated that HIV came from outer space. After Hoyle's death, The Lancet published a letter to the editor from Wickramasinghe and two of his colleagues, in which they hypothesized that the virus that causes severe acute respiratory syndrome (SARS) could be extraterrestrial in origin and not originated from chickens. The Lancet subsequently published three responses to this letter, showing that the hypothesis was not evidence-based, and casting doubts on the quality of the experiments referenced by Wickramasinghe in his letter. A 2008 encyclopedia notes that "Like other claims linking terrestrial disease to extraterrestrial pathogens, this proposal was rejected by the greater research community."

In April 2016, Jiangwen Qu of the Department of Infectious Disease Control in China presented a statistical study suggesting that "extremes of sunspot activity to within plus or minus 1 year may precipitate influenza pandemics." He discussed possible mechanisms of epidemic initiation and early spread, including speculation on primary causation by externally derived viral variants from space via cometary dust.

Case studies

• A meteorite originating from Mars known as ALH84001 was shown
in 1996 to contain microscopic structures resembling small terrestrial
nanobacteria. When the discovery was announced, many immediately
conjectured that these were fossils and were the first evidence of extraterrestrial life — making headlines around the world. Public interest soon
started to dwindle as most experts started to agree that these structures
were not indicative of life, but could instead be formed abiotically from
organic molecules. However, in November 2009, a team of scientists at
Johnson Space Center, including David McKay, reasserted that there was
"strong evidence that life may have existed on ancient Mars", after having
reexamined the meteorite and finding magnetite crystals.

• On May 11, 2001, two researchers from the University of Naples claimed
to have found viable extraterrestrial bacteria inside a meteorite. Geolo-
gist Bruno D'Argenio and molecular biologist Giuseppe Geraci claim the
bacteria were wedged inside the crystal structure of minerals, but were
resurrected when a sample of the rock was placed in a culture medium.

• An Indian and British team of researchers led by Chandra Wickramasinghe reported on 2001 that air samples over Hyderabad, India, gathered
from the stratosphere by the Indian Space Research Organisation (ISRO)
on Jan 21, 2001, contained clumps of living cells. Wickramasinghe calls
this "unambiguous evidence for the presence of clumps of living cells in
air samples from as high as 41 km, above which no air from lower down
would normally be transported". Two bacterial and one fungal species
were later independently isolated from these filters which were identified
as *Bacillus simplex*, *Staphylococcus pasteuri* and *Engyodontium album*
respectively. Pushkar Ganesh Vaidya from the Indian Astrobiology Research Centre reported in 2009 that "the three microorganisms captured
during the balloon experiment do not exhibit any distinct adaptations
expected to be seen in microorganisms occupying a cometary niche".

• In 2005 an improved experiment was conducted by ISRO. On April 20,
2005, air samples were collected from the upper atmosphere at altitudes
ranging from 20 km to more than 40 km. The samples were tested at two
labs in India. The labs found 12 bacterial and 6 different fungal species
in these samples. The fungi were *Penicillium decumbens*, *Cladosporium
cladosporioides*, *Alternaria sp.* and *Tilletiopsis albescens*. Out of the 12
bacterial samples, three were identified as new species and named *Jani-
bacter hoylei* (after Fred Hoyle), *Bacillus isronensis* (named after ISRO)
and *Bacillus aryabhattai* (named after the ancient Indian mathematician, Aryabhata). These three new species showed that they were more
resistant to UV radiation than similar bacteria.[303]

Some other researchers have retrieved bacteria from the stratosphere since the 1970s.[304] Atmospheric sampling by NASA in 2010 before and after hurricanes, collected 314 different types of bacteria; the study suggests that large-scale convection during tropical storms and hurricanes can then carry this material from the surface higher up into the atmosphere.

- Another proposed mechanism of spores in the stratosphere is lifting by weather and Earth magnetism up to the ionosphere into low Earth orbit, where Russian astronauts retrieved DNA from a known sterile exterior surface of the International Space Station.[305] The Russian scientists then also speculated the possibility "that common terrestrial bacteria are constantly being resupplied from space."

- On January 10, 2013, Chandra Wickramasinghe found fossil diatom frustules in what he thinks is a new kind of carbonaceous meteorite called Polonnaruwa that landed in the North Central Province of Sri Lanka on 29 December 2012. Early on, there was criticism that Wickramasinghe's report was not an examination of an actual meteorite but of some terrestrial rock passed off as a meteorite.

Wickramasinghe's team remark that they are aware that a large number of unrelated stones have been submitted for analysis, and have no knowledge regarding the nature, source or origin of the stones their critics have examined, so Wickramasinghe clarifies that he is using the stones submitted by the Medical Research Institute in Sri Lanka. In response to the criticism from other scientists, Wickramasinghe performed X-ray diffraction and isotope analyses to verify its meteoritic origin. His analysis revealed a 95% silica and 3% quartz content, and interpreted this result as a "carbonaceous meteorite of unknown type". In addition, Wickramasinghe's team remarked that the temperature at which sand must be heated by lightning to melt and form a fulgurite (1770 °C) would have vaporized and burned all carbon-rich organisms and melted and thus destroyed the delicately marked silica frustules of the diatoms, and that the oxygen isotope data confirms its meteoric origin. Wickramasinghe's team also argues that since living diatoms require nitrogen fixation to synthetize amino acids, proteins, DNA, RNA and other life-critical biomolecules, a population of extraterrestrial cyanobacteria must have been a required component of the comet (Polonnaruwa meteorite) "ecosystem".

- In 2013, Dale Warren Griffin, a microbiologist working at the United States Geological Survey noted that viruses are the most numerous entities on Earth. Griffin speculates that viruses evolved in comets and on other planets and moons may be pathogenic to humans, so he proposed to also look for viruses on moons and planets of the Solar System.

Figure 57: *Hydrothermal vents are able to support extremophile bacteria on Earth and may also support life in other parts of the cosmos.*

Hoaxes

A separate fragment of the Orgueil meteorite (kept in a sealed glass jar since its discovery) was found in 1965 to have a seed capsule embedded in it, whilst the original glassy layer on the outside remained undisturbed. Despite great initial excitement, the seed was found to be that of a European Juncaceae or Rush plant that had been glued into the fragment and camouflaged using coal dust. The outer "fusion layer" was in fact glue. Whilst the perpetrator of this hoax is unknown, it is thought that they sought to influence the 19th century debate on spontaneous generation — rather than panspermia — by demonstrating the transformation of inorganic to biological matter.

Extremophiles

Until the 1970s, life was thought to depend on its access to sunlight. Even life in the ocean depths, where sunlight cannot reach, was believed to obtain its nourishment either from consuming organic detritus rained down from the surface waters or from eating animals that did. However, in 1977, during an exploratory dive to the Galapagos Rift in the deep-sea exploration submersible *Alvin*, scientists discovered colonies of assorted creatures clustered around undersea volcanic features known as black smokers. It was soon determined that the basis for this food chain is a form of bacterium that derives its energy from oxidation of reactive chemicals, such as hydrogen or hydrogen sulfide, that bubble up from the Earth's interior. This chemosynthesis revolutionized the study of biology by revealing that terrestrial life need not be Sun-dependent; it only requires water and an energy gradient in order to exist.

It is now known that extremophiles, microorganisms with extraordinary capability to thrive in the harshest environments on Earth, can specialize to thrive in the deep-sea, ice, boiling water, acid, the water core of nuclear reactors, salt crystals, toxic waste and in a range of other extreme habitats that were previously thought to be inhospitable for life.[306] Living bacteria found in ice core samples retrieved from 3,700 metres (12,100 ft) deep at Lake Vostok in Antarctica, have provided data for extrapolations to the likelihood of microorganisms surviving frozen in extraterrestrial habitats or during interplanetary transport. Also, bacteria have been discovered living within warm rock deep in the Earth's crust.

In order to test some these organisms' potential resilience in outer space, plant seeds and spores of bacteria, fungi and ferns have been exposed to the harsh space environment. Spores are produced as part of the normal life cycle of many plants, algae, fungi and some protozoans, and some bacteria produce endospores or cysts during times of stress. These structures may be highly resilient to ultraviolet and gamma radiation, desiccation, lysozyme, temperature, starvation and chemical disinfectants, while metabolically inactive. Spores germinate when favourable conditions are restored after exposure to conditions fatal to the parent organism.

Although computer models suggest that a captured meteoroid would typically take some tens of millions of years before collision with a planet, there are documented viable Earthly bacterial spores that are 40 million years old that are very resistant to radiation, and others able to resume life after being dormant for 25 million years,[307] suggesting that lithopanspermia life-transfers are possible via meteorites exceeding 1 m in size.

The discovery of deep-sea ecosystems, along with advancements in the fields of astrobiology, observational astronomy and discovery of large varieties of extremophiles, opened up a new avenue in astrobiology by massively expanding the number of possible extraterrestrial habitats and possible transport of hardy microbial life through vast distances.

Figure 58: *EURECA facility deployment in 1992*

Research in outer space

The question of whether certain microorganisms can survive in the harsh environment of outer space has intrigued biologists since the beginning of spaceflight, and opportunities were provided to expose samples to space. The first American tests were made in 1966, during the Gemini IX and XII missions, when samples of bacteriophage T1 and spores of *Penicillium roqueforti* were exposed to outer space for 16.8 h and 6.5 h, respectively. Other basic life sciences research in low Earth orbit started in 1966 with the Soviet biosatellite program Bion and the U.S. Biosatellite program. Thus, the plausibility of panspermia can be evaluated by examining life forms on Earth for their capacity to survive in space. The following experiments carried on low Earth orbit specifically tested some aspects of panspermia or lithopanspermia:

ERA

The Exobiology Radiation Assembly (ERA) was a 1992 experiment on board the European Retrievable Carrier (EURECA) on the biological effects of space radiation. EURECA was an unmanned 4.5 tonne satellite with a payload of 15 experiments. It was an astrobiology mission developed by the European Space Agency (ESA). Spores of different strains of *Bacillus subtilis* and the *Escherichia coli* plasmid pUC19 were exposed to selected conditions of space (space vacuum and/or defined wavebands and intensities of solar ultraviolet radiation). After the approximately 11-month mission, their responses were studied in terms of survival, mutagenesis in the *his* (*B. subtilis*) or *lac* locus (pUC19), induction of DNA strand breaks, efficiency of DNA repair systems, and the role of external protective agents. The data were compared with those of a simultaneously running ground control experiment:

*   The survival of spores treated with the vacuum of space, however shielded against solar radiation, is substantially increased, if they are exposed in multilayers and/or in the presence of glucose as protective.
*   All spores in "artificial meteorites", i.e. embedded in clays or simulated Martian soil, are killed.
*   Vacuum treatment leads to an increase of mutation frequency in spores, but not in plasmid DNA.
*   Extraterrestrial solar ultraviolet radiation is mutagenic, induces strand breaks in the DNA and reduces survival substantially.
*   Action spectroscopy confirms results of previous space experiments of a synergistic action of space vacuum and solar UV radiation with DNA being the critical target.
*   The decrease in viability of the microorganisms could be correlated with the increase in DNA damage.
*   The purple membranes, amino acids and urea were not measurably affected by the dehydrating condition of open space, if sheltered from solar radiation. Plasmid DNA, however, suffered a significant amount of strand breaks under these conditions.

BIOPAN

BIOPAN is a multi-user experimental facility installed on the external surface of the Russian Foton descent capsule. Experiments developed for BIOPAN are designed to investigate the effect of the space environment on biological material after exposure between 13 and 17 days. The experiments in BIOPAN are exposed to solar and cosmic radiation, the space vacuum and weightlessness, or a selection thereof. Of the 6 missions flown so far on BIOPAN between 1992 and 2007, dozens of experiments were conducted, and some analyzed the likelihood of panspermia. Some bacteria, lichens (*Xanthoria elegans*, *Rhizocarpon geographicum* and their mycobiont cultures, the black Antarctic microfungi *Cryomyces minteri* and *Cryomyces antarcticus*), spores, and even one animal (tardigrades) were found to have survived the harsh outer space environment and cosmic radiation.

Figure 59: *EXOSTACK on the Long Duration Exposure Facility satellite.*

EXOSTACK

The German EXOSTACK experiment was deployed on 7 April 1984 on board the Long Duration Exposure Facility statellite. 30% of *Bacillus subtilis* spores survived the nearly 6 years exposure when embedded in salt crystals, whereas 80% survived in the presence of glucose, which stabilize the structure of the cellular macromolecules, especially during vacuum-induced dehydration.

If shielded against solar UV, spores of *B. subtilis* were capable of surviving in space for up to 6 years, especially if embedded in clay or meteorite powder (artificial meteorites). The data support the likelihood of interplanetary transfer of microorganisms within meteorites, the so-called lithopanspermia hypothesis.

EXPOSE

EXPOSE is a multi-user facility mounted outside the International Space Station dedicated to astrobiology experiments. There have been three EXPOSE experiments flown between 2008 and 2015: EXPOSE-E, EXPOSE-R and EXPOSE-R2.

Results from the orbital missions, especially the experiments *SEEDS* and *LiFE*, concluded that after an 18-month exposure, some seeds and lichens

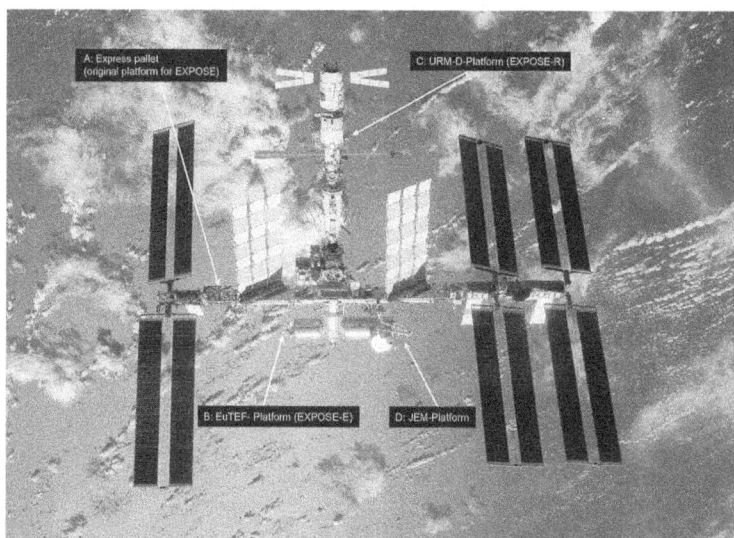

Figure 60: *Location of the astrobiology EXPOSE-E and EXPOSE-R facilities on the International Space Station*

(*Stichococcus sp.* and *Acarospora sp.*, a lichenized fungal genus) may be capable to survive interplanetary travel if sheltered inside comets or rocks from cosmic radiation and UV radiation. The *LIFE, SPORES,* and *SEEDS* parts of the experiments provided information about the likelihood of lithopanspermia. These studies will provide experimental data to the lithopanspermia hypothesis, and they will provide basic data to planetary protection issues.

Tanpopo

The Tanpopo mission is an orbital astrobiology experiment by Japan that is currently investigating the possible interplanetary transfer of life, organic compounds, and possible terrestrial particles in low Earth orbit. The Tanpopo experiment is taking place at the Exposed Facility located on the exterior of Kibo module of the International Space Station. The mission will collect cosmic dusts and other particles for three years by using an ultra-low density silica gel called aerogel. The purpose is to assess the panspermia hypothesis and the possibility of natural interplanetary transport of life and its precursors.[308,309] Some of these aerogels will be replaced every one or two years through 2018.[310] Sample collection began in May 2015, and the first samples were be returned to Earth in mid-2016.

Figure 61: *Dust collector with aerogel blocks*

Criticism

Panspermia is often criticized because it does not answer the question of the origin of life but merely places it on another celestial body. It was also criticized because it was thought it could not be tested experimentally.

Wallis and Wickramasinghe argued in 2004 that the transport of individual bacteria or clumps of bacteria, is overwhelmingly more important than lithopanspermia in terms of numbers of microbes transferred, even accounting for the death rate of unprotected bacteria in transit. Then it was found that isolated spores of *B. subtilis* were killed by several orders of magnitude if exposed to the full space environment for a mere few seconds. These results clearly negate the original panspermia hypothesis, which requires single spores as space travelers accelerated by the radiation pressure of the Sun, requiring many years to travel between the planets. However, if shielded against solar UV, spores of *Bacillus subtilis* were capable of surviving in space for up to 6 years, especially if embedded in clay or meteorite powder (artificial meteorites). The data support the likelihood of interplanetary transfer of microorganisms within meteorites, the so-called **lithopanspermia** hypothesis.

Further reading

- Crick, F (1981), *Life, Its Origin and Nature*, Simon & Schuster, ISBN 0-7088-2235-5.
- Hoyle, F (1983), *The Intelligent Universe*, London: Michael Joseph, ISBN 0-7181-2298-4.

External links

シ ひ Ꮍ え ℓ↲ ᑌ ᵕ ᔦ ᓍ	Look up *panspermia* in Wiktionary, the free dictionary.

- A.E. Zlobin, 2013, Tunguska similar impacts and origin of life (mathematical theory of origin of life; incoming of pattern recognition algorithm due to comets)[311]
- Francis Crick's notes[312] for a lecture on directed panspermia, dated 5 November 1976.
- "Earth sows its seeds in space". *Nature News*. 23 February 2004. doi: 10.1038/news040216-20[313] (inactive 2017-01-14).
- Warmflash, D.; Weiss, B. (24 October 2005). "Did Life Come from Another World?"[314]. *Scientific American*. **293** (5): 64–71. Bibcode: 2005SciAm.293e..64W[315]. doi: 10.1038/scientificamerican1105-64[316].

Extraterrestrial organic molecules

List of interstellar and circumstellar molecules

This is a list of molecules that have been detected in the interstellar medium and circumstellar envelopes, grouped by the number of component atoms. The chemical formula is listed for each detected compound, along with any ionized form that has also been observed.

Detection

The molecules listed below were detected by spectroscopy. Their spectral features are generated by transitions of component electrons between different energy levels, or by rotational or vibrational spectra. Detection usually occurs in radio, microwave, or infrared portions of the spectrum.

Interstellar molecules are formed by chemical reactions within very sparse interstellar or circumstellar clouds of dust and gas. Usually this occurs when a molecule becomes ionized, often as the result of an interaction with a cosmic ray. This positively charged molecule then draws in a nearby reactant by electrostatic attraction of the neutral molecule's electrons. Molecules can also be generated by reactions between neutral atoms and molecules, although this process is generally slower. The dust plays a critical role of shielding the molecules from the ionizing effect of ultraviolet radiation emitted by stars.

History

The chemistry of life may have begun shortly after the Big Bang, 13.8 billion years ago, during a habitable epoch when the Universe was only 10–17 million years old.

The first carbon-containing molecule detected in the interstellar medium was the methylidyne radical (CH·) in 1937. From the early 1970s it was becoming evident that interstellar dust consisted of a large component of more complex organic molecules (COMs), probably polymers. Chandra Wickramasinghe proposed the existence of polymeric composition based on the molecule formaldehyde (H_2CO).[317] Fred Hoyle and Chandra Wickramasinghe later proposed the identification of bicyclic aromatic compounds from an analysis of the ultraviolet extinction absorption at 2175 Å,[318] thus demonstrating the existence of polycyclic aromatic hydrocarbon molecules in space.

In 2004, scientists reported detecting the spectral signatures of anthracene and pyrene in the ultraviolet light emitted by the Red Rectangle nebula (no other such complex molecules had ever been found before in outer space). This discovery was considered a confirmation of a hypothesis that as nebulae of the same type as the Red Rectangle approach the ends of their lives, convection currents cause carbon and hydrogen in the nebulae's core to get caught in stellar winds, and radiate outward. As they cool, the atoms supposedly bond to each other in various ways and eventually form particles of a million or more atoms. The scientists inferred that since they discovered polycyclic aromatic hydrocarbons (PAHs) — which may have been vital in the formation of early life on Earth — in a nebula, by necessity they must originate in nebulae.

In 2010, fullerenes (or "buckyballs") were detected in nebulae. Fullerenes have been implicated in the origin of life; according to astronomer Letizia Stanghellini, "It's possible that buckyballs from outer space provided seeds for life on Earth."

In October 2011, scientists found using spectroscopy that cosmic dust contains complex organic compounds ("amorphous organic solids with a mixed aromatic-aliphatic structure") that could be created naturally, and rapidly, by stars. The compounds are so complex that their chemical structures resemble the makeup of coal and petroleum; such chemical complexity was previously thought to arise only from living organisms. These observations suggest that organic compounds introduced on Earth by interstellar dust particles could serve as basic ingredients for life due to their surface-catalytic activities. One of the scientists suggested that these compounds may have been related to the development of life on Earth and said that, "If this is the case, life on Earth may have had an easier time getting started as these organics can serve as basic ingredients for life."

In August 2012, astronomers at Copenhagen University reported the detection of a specific sugar molecule, glycolaldehyde, in a distant star system. The molecule was found around the protostellar binary *IRAS 16293-2422*, which is located 400 light years from Earth. Glycolaldehyde is needed to form ribonucleic acid, or RNA, which is similar in function to DNA. This finding suggests that complex organic molecules may form in stellar systems prior to the formation of planets, eventually arriving on young planets early in their formation.

In September 2012, NASA scientists reported that PAHs, subjected to interstellar medium (ISM) conditions, are transformed, through hydrogenation, oxygenation, and hydroxylation, to more complex organics — "a step along the path toward amino acids and nucleotides, the raw materials of proteins and DNA, respectively". Further, as a result of these transformations, the PAHs lose their spectroscopic signature which could be one of the reasons "for the lack of PAH detection in interstellar ice grains, particularly the outer regions of cold, dense clouds or the upper molecular layers of protoplanetary disks."

PAHs are found everywhere in deep space and, in June 2013, PAHs were detected in the upper atmosphere of Titan, the largest moon of the planet Saturn.

In 2013, Dwayne Heard at the University of Leeds suggested[319] that quantum mechanical tunneling could explain a reaction his group observed taking place, at a significantly higher than expected rate, between cold (around 63 kelvins) hydroxyl and methanol molecules, apparently bypassing intramolecular energy barriers which would have to be overcome by thermal energy or ionization events for the same rate to exist at warmer temperatures. The proposed tunneling mechanism may help explain the common observation of fairly complex molecules (up to tens of atoms) in interstellar space.

A particularly large and rich region for detecting interstellar molecules is Sagittarius B2 (Sgr B2). This giant molecular cloud lies near the center of the Milky Way galaxy and is a frequent target for new searches. About half of the molecules listed below were first found near Sgr B2, and nearly every other molecule has since been detected in this feature. A rich source of investigation for circumstellar molecules is the relatively nearby star CW Leonis (IRC +10216), where about 50 compounds have been identified.

In March 2015, NASA scientists reported that, for the first time, complex DNA and RNA organic compounds of life, including uracil, cytosine and thymine, have been formed in the laboratory under outer space conditions, using starting chemicals, such as pyrimidine, found in meteorites. Pyrimidine, like polycyclic aromatic hydrocarbons (PAHs), the most carbon-rich chemical found in the Universe, may have been formed in red giants or in interstellar dust and gas clouds, according to the scientists.

In October 2016, astronomers reported that the very basic chemical ingredients of life—the carbon-hydrogen molecule (CH', or methylidyne radical), the carbon-hydrogen positive ion (CH^+ cation) and the carbon ion (C^+ cation)—are the result, in large part, of ultraviolet light from stars, rather than in other ways, such as the result of turbulent events related to supernovae and young stars, as thought earlier.

Theoretical models

To explain the observed ratios of isomeric compounds, the minimum energy principle has been used. In the majority of cases, it explains that some organic entities have greater abundance than their isomers due to the lower total energies of the first one. However, a few exceptions where the principle fails are also known.

Another approach ignores energy and deals only with the molecular complexity estimated by the information entropy index. It speculates that the points of several natural compounds (urea, pyrimidine, dihydroxyacetone, uracil, cytosine, glycine, and alanine) fall into the range of the values typical for the known interstellar molecules that indicates high probability of their detection in interstellar environment. Additionally the molecules with maximal information entropy, i.e. the most complex compounds, make up approximately a half of the interstellar set and their percentage is decreased with the size. This trend may be associated with the different stabilities of the molecules with uniform (usually more stable) and diversified (usually less stable) chemical structures, so the detectable molecules with a large size must possess symmetric structure more probably than non-symmetric. The remarkable detection of low-entropy (highly symmetric) fullerene molecules supports this assumption. It is also noted that information entropy reflects the depth of hydrogenation of interstellar entities: the molecules with maximal information entropy are hydrogen-poor whereas the others are mainly hydrogen-rich.

Molecules

The following tables list molecules that have been detected in the interstellar medium, grouped by the number of component atoms. If there is no entry in the molecule column, only the ionized form has been detected. For molecules where no designation was given in the scientific literature, that field is left empty. Mass is given in atomic mass units. The total number of unique species, including distinct ionization states, is listed in parentheses in each section header.

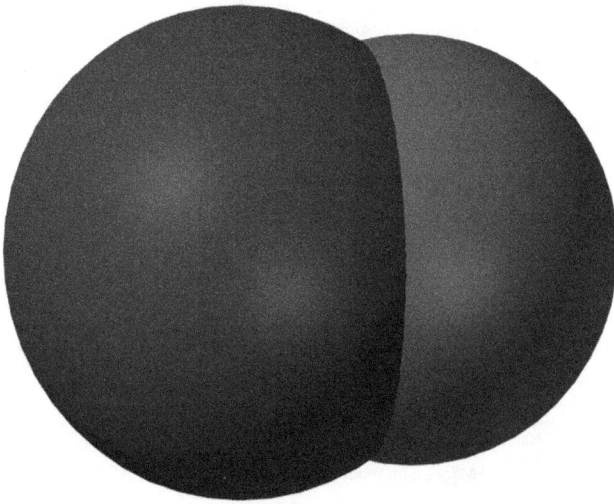

Figure 62: *Carbon monoxide is frequently used to trace the distribution of mass in molecular clouds.*

Most of the molecules detected so far are organic. Only one inorganic species has been observed in molecules which contain at least five atoms, SiH_4. Larger molecules have so far all had at least one carbon atom, with no N–N or O–O bonds.

Diatomic (43)

Molecule	Designation	Mass	Ions
AlCl	Aluminium monochloride	62.5	—
AlF	Aluminium monofluoride	46	—
AlO	Aluminium monoxide	43	—
—	Argonium	41	ArH+
C_2	Diatomic carbon	24	—
—	Fluoromethylidynium	31	CF+
CH	Methylidyne radical	13	CH+
CN	Cyanogen radical	26	CN+, CN−
CO	Carbon monoxide	28	CO+

CP	Carbon monophosphide	43	—
CS	Carbon monosulfide	44	—
FeO	Iron(II) oxide	82	—
H_2	Molecular hydrogen	2	—
HCl	Hydrogen chloride	36.5	HCl^+
HF	Hydrogen fluoride	20	—
HO	Hydroxyl radical	17	OH^+
KCl	Potassium chloride	75.5	—
NH	Nitrogen monohydride	15	—
N_2	Molecular nitrogen	28	—
NO	Nitric oxide	30	NO^+
NS	Nitrogen sulfide	46	—
NaCl	Sodium chloride	58.5	—
—	Magnesium monohydride cation	25.3	MgH^+
NaI	Sodium iodide	150	—
O_2	Molecular oxygen	32	—
PN	Phosphorus mononitride	45	—
PO	Phosphorus monoxide	47	—
SH	Sulfur monohydride	33	SH^+
SO	Sulfur monoxide	48	SO^+
SiC	Carborundum	40	—
SiN	Silicon mononitride	42	—
SiO	Silicon monoxide	44	—
SiS	Silicon monosulfide	60	—
TiO	Titanium oxide	63.9	—

Triatomic (41)

Molecule	Designation	Mass	Ions
AlNC	Aluminium isocyanide	53	—
AlOH	Aluminium hydroxide	44	—
C_3	Tricarbon	36	—
C_2H	Ethynyl radical	25	—
CCN	Cyanomethylidyne	38	—
C_2O	Dicarbon monoxide	40	—
C_2S	Thioxoethenylidene	56	—

C_2P	—	55	—
CO_2	Carbon dioxide	44	—
FeCN	Iron cyanide	82	—
—	Protonated molecular hydrogen	3	H_3^+
H_2C	Methylene radical	14	—
—	Chloronium	37.5	H_2Cl^+
H_2O	Water	18	H_2O^+
HO_2	Hydroperoxyl	33	—
H_2S	Hydrogen sulfide	34	—
HCN	Hydrogen cyanide	27	—
HNC	Hydrogen isocyanide	27	—
HCO	Formyl radical	29	HCO^+
HCP	Phosphaethyne	44	—
HCS	Thioformyl	45	HCS^+
—	Diazenylium	29	HN_2^+
HNO	Nitroxyl	31	—
—	Isoformyl	29	HOC^+
HSC	Isothioformyl	45	—
KCN	Potassium cyanide	65	—
MgCN	Magnesium cyanide	50	—
MgNC	Magnesium isocyanide	50	—
NH_2	Amino radical	16	—
N_2O	Nitrous oxide	44	—
NaCN	Sodium cyanide	49	—
NaOH	Sodium hydroxide	40	—
OCS	Carbonyl sulfide	60	—
O_3	Ozone	48	—
SO_2	Sulfur dioxide	64	—
c-SiC_2	c-Silicon dicarbide	52	—
SiCSi	Disilicon carbide	68	—
SiCN	Silicon carbonitride	54	—
SiNC		54	—
TiO_2	Titanium dioxide	79.9	—

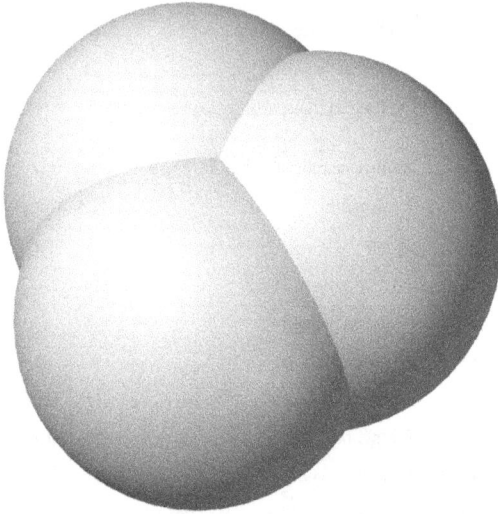

Figure 63: *The H$^+$*
$_3$ cation is one of the most abundant ions in
the universe. It was first detected in 1993.

Four atoms (28)

Molecule	Designation	Mass	Ions
CH_3	Methyl radical	15	—
l-C_3H	Propynylidyne	37	l-C_3H^+
c-C_3H	Cyclopropynylidyne	37	—
C_3N	Cyanoethynyl	50	C_3N^-
C_3O	Tricarbon monoxide	52	—
C_3S	Tricarbon sulfide	68	—
—	Hydronium	19	H_3O^+
C_2H_2	Acetylene	26	—
H_2CN	Methylene amidogen	28	H_2CN^+
H_2CO	Formaldehyde	30	—
H_2CS	Thioformaldehyde	46	—
HCCN	—	39	—
HCCO	Ketenyl	41	—

—	Protonated hydrogen cyanide	28	$HCNH^+$
—	Protonated carbon dioxide	45	$HOCO^+$
HCNO	Fulminic acid	43	—
HOCN	Cyanic acid	43	—
CNCN	Isocyanogen	52	—
HOOH	Hydrogen peroxide	34	—
HNCO	Isocyanic acid	43	—
HNCS	Isothiocyanic acid	59	—
NH_3	Ammonia	17	—
HSCN	Thiocyanic acid	59	—
SiC_3	Silicon tricarbide	64	—
HMgNC	Hydromagnesium isocyanide	51.3	—

Five atoms (20)

Molecule	Designation	Mass	Ions
—	Ammonium ion[320,321]	18	NH_4^+
CH_4	Methane	16	—
CH_3O	Methoxy radical	31	—
$c\text{-}C_3H_2$	Cyclopropenylidene	38	—
$l\text{-}H_2C_3$	Propadienylidene	38	—
H_2CCN	Cyanomethyl	40	—
H_2C_2O	Ketene	42	—
H_2CNH	Methylenimine	29	—
HNCNH	Carbodiimide	42	—
—	Protonated formaldehyde	31	H_2COH^+
C_4H	Butadiynyl	49	C_4H^-
HC_3N	Cyanoacetylene	51	—
HCC-NC	Isocyanoacetylene	51	—
HCOOH	Formic acid	46	—
NH_2CN	Cyanamide	42	—
—	Protonated cyanogen	53	$NCCNH^+$
HC(O)CN	Cyanoformaldehyde	55	—
C_5	Linear C_5	60	—
SiC_4	Silicon-carbide cluster	92	—

| SiH_4 | Silane | 32 | — |

Six atoms (16)

Molecule	Designation	Mass	Ions
c-H_2C_3O	Cyclopropenone	54	—
E-HNCHCN	E-Cyanomethanimine	54	—
C_2H_4	Ethylene	28	—
CH_3CN	Acetonitrile	40	—
CH_3NC	Methyl isocyanide	40	—
CH_3OH	Methanol[322]	32	—
CH_3SH	Methanethiol	48	—
l-H_2C_4	Diacetylene	50	—
—	Protonated cyanoacetylene	52	HC_3NH^+
$HCONH_2$	Formamide	44	—
C_5H	Pentynylidyne	61	—
C_5N	Cyanobutadiynyl radical	74	—
HC_2CHO	Propynal	54	—
HC_4N	—	63	—
CH_2CNH	Ketenimine	40	—
C_5S	—	92	—

Seven atoms (11)

Molecule	Designation	Mass	Ions
c-C_2H_4O	Ethylene oxide	44	—
CH_3C_2H	Methylacetylene	40	—
H_3CNH_2	Methylamine	31	—
CH_2CHCN	Acrylonitrile	53	—
H_2CHCOH	Vinyl alcohol	44	—
C_6H	Hexatriynyl radical	73	C_6H^-
HC_4CN	Cyanodiacetylene	75	—
HC_5O	—	77	—
CH_3CHO	Acetaldehyde	44	—
CH_3NCO	Methyl isocyanate	57	—

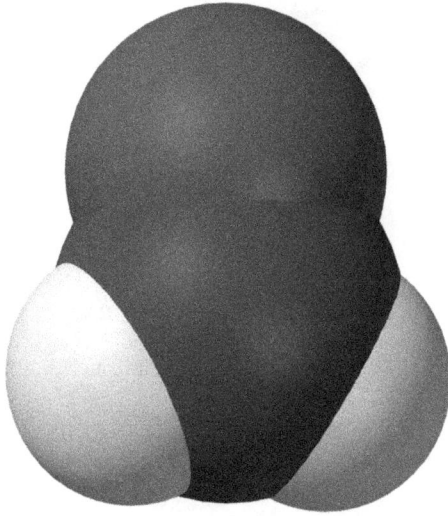

Figure 64: *Formaldehyde is an organic molecule that is widely distributed in the interstellar medium.*

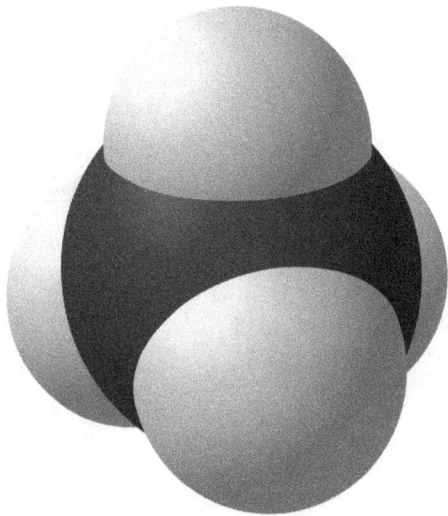

Figure 65: *Methane, the primary component of natural gas, has also been detected on comets and in the atmosphere of several planets in the Solar System.*

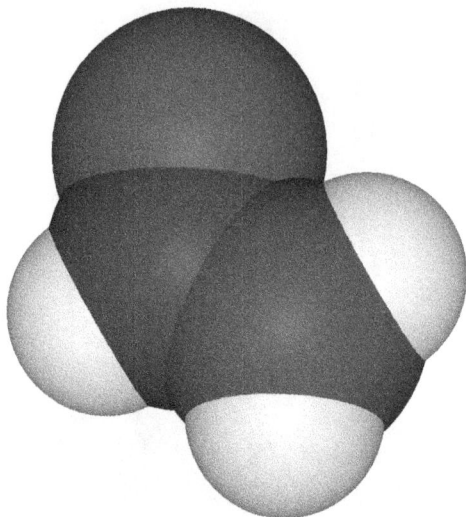

Figure 66: *In the ISM, formamide (above) can combine with methylene to form acetamide.*

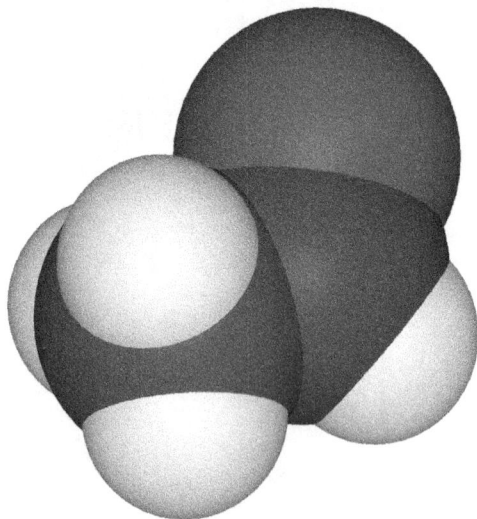

Figure 67: *Acetaldehyde (above) and its isomers vinyl alcohol and ethylene oxide have all been detected in interstellar space.*

Figure 68: *The radio signature of acetic acid, a compound found in vinegar, was confirmed in 1997.*

Eight atoms (11)

Molecule	Designation	Mass
H_3CC_2CN	Methylcyanoacetylene	65
$H_2COHCHO$	Glycolaldehyde	60
$HCOOCH_3$	Methyl formate	60
CH_3COOH	Acetic acid	60
H_2C_6	Hexapentaenylidene	74
CH_2CHCHO	Propenal	56
CH_2CCHCN	Cyanoallene	65
CH_3CHNH	Ethanimine	43
C_7H	Heptatrienyl radical	85
NH_2CH_2CN	Aminoacetonitrile	56
$(NH_2)_2CO$	Urea	60

Nine atoms (10)

Molecule	Designation	Mass	Ions
CH_3C_4H	Methyldiacetylene	64	—
CH_3OCH_3	Dimethyl Ether	46	—
CH_3CH_2CN	Propionitrile	55	—
CH_3CONH_2	Acetamide	59	—
CH_3CH_2OH	Ethanol	46	—
C_8H	Octatetraynyl radical	97	C_8H^-
HC_7N	Cyanohexatriyne or Cyanotriacetylene	99	—
CH_3CHCH_2	Propylene (propene)	42	—
CH_3CH_2SH	Ethyl mercaptan	62	—

A number of polyyne-derived chemicals are among the heaviest molecules found in the interstellar medium.

Ten or more atoms (17)

Atoms	Molecule	Designation	Mass	Ions
10	$(CH_3)_2CO$	Acetone	58	—
10	$(CH_2OH)_2$	Ethylene glycol	62	—
10	CH_3CH_2CHO	Propanal	58	—
10	CH_3OCH_2OH	Methoxymethanol	62	—
10	CH_3C_5N	Methyl-cyano-diacetylene	89	—

10	CH_3CHCH_2O	Propylene oxide	58	—
11	HC_8CN	Cyanotetra-acetylene	123	—
11	C_2H_5OCHO	Ethyl formate	74	—
11	CH_3COOCH_3	Methyl acetate	74	—
11	CH_3C_6H	Methyltriacetylene	88	—
12	C_6H_6	Benzene	78	—
12	C_3H_7CN	n-Propyl cyanide	69	—
12	$(CH_3)_2CHCN$	iso-Propyl cyanide	69	—
13	C_6H_5CN	Benzonitrile	104	—
13	$HC_{10}CN$	Cyanopentaacetylene	147	—
60	C_{60}	Buckminsterfullerene (C_{60} fullerene)	720	C_{60}^+
70	C_{70}	C_{70} fullerene	840	—

Deuterated molecules (20)

These molecules all contain one or more deuterium atoms, a heavier isotope of hydrogen.

Atoms	Molecule	Designation
2	HD	Hydrogen deuteride
3	H_2D^+, HD_2^+	Trihydrogen cation
3	HDO, D_2O	Heavy water
3	DCN	Hydrogen cyanide
3	DCO	Formyl radical
3	DNC	Hydrogen isocyanide
3	N_2D^+	—
4	NH_2D, NHD_2, ND_3	Ammonia
4	HDCO, D_2CO	Formaldehyde
4	DNCO	Isocyanic acid
5	NH_3D^+	Ammonium ion
6	NH_2CDO; NHDCHO	Formamide
7	CH_2DCCH, CH_3CCD	Methylacetylene

Unconfirmed (12)

Evidence for the existence of the following molecules has been reported in scientific literature, but the detections are either described as tentative by the authors, or have been challenged by other researchers. They await independent confirmation.

Atoms	Molecule	Designation
2	SiH	Silylidine
4	PH_3	Phosphine
4	MgCCH	Magnesium monoacetylide
4	NCCP	Cyanophosphaethyne
5	H_2NCO^+	—
4	SiH_3CN	Silyl cyanide
10	H_2NH_2CCOOH	Glycine
12	$CO(CH_2OH)_2$	Dihydroxyacetone
12	$C_2H_5OCH_3$	Ethyl methyl ether
18	$C_{10}H^+_8$	Naphthalene cation
24	C_{24}	Graphene
24	$C_{14}H_{10}$	Anthracene
26	$C_{16}H_{10}$	Pyrene

External links

- Woon, David E. (October 1, 2010). "Interstellar and Circumstellar Molecules"[323]. Retrieved 2010-10-04.
- "Molecules in Space"[324]. Universität zu Köln. August 2010. Retrieved 2010-10-04.
- Dworkin, Jason P. (February 1, 2007). "Interstellar Molecules"[325]. NASA's Cosmic Ice Lab. Retrieved 2010-12-23.
- Wootten, Al (November 2005). "The 129 reported interstellar and circumstellar molecules"[326]. National Radio Astronomy Observatory. Retrieved 2007-02-13.
- Lovas, F. J.; Dragoset, R. A. (February 2004). "NIST Recommended Rest Frequencies for Observed Interstellar Molecular Microwave Transitions, 2002 Revision"[327]. National Institute of Standards and Technology. Retrieved 2007-02-13.

<indicator name="featured-star"> ★ </indicator>

Lipid world

Gard model

In evolutionary biology, the **GARD (Graded Autocatalysis Replication Domain) model** is a general kinetic model for homeostatic-growth and fission of compositional-assemblies, with specific application towards lipids.

In the context of abiogenesis, the lipid-world suggests assemblies of simple molecules, such as lipids, can store and propagate information, thus undergo evolution.

These 'compositional assemblies' have been suggested to play a role in the origin of life. The idea is the information being transferred throughout the generations is *compositional information* – the different types and quantities of molecules within an assembly. This is different from the information encoded in RNA or DNA, which is the specific sequence of bases in such molecule. Thus, the model is viewed as an alternative or an ancestor to the RNA world hypothesis.

The model

The composition vector of an assembly is written as: $v = n_1 \cdots n_{N_G}$. Where $n_1 \cdots n_{N_G}$ are the molecular counts of lipid type i within the assembly, and NG is how many different lipid types exist (*repertiore size*).

The change in the count of molecule type i is described by:

$$\frac{dn_i}{dt} = (k_f \rho_i N - k_b n_i) \left(1 + \sum_{j=1}^{N_G} \beta_{ij} \frac{n_j}{N} \right)$$

k_f and k_b are the basel forward (joining) and backward (leaving) rate constants, β_{ij} is a non-negative rate enhancement exerted by molecule type j within the assembly on type i from the environment, and ρ is the environmental concentration of each molecule type. β is viewed as a directed, weighted, complex network.

The assembly current size is $N = \sum_{i=1}^{N_G} n_i$. The system is kept away from equilibrium by imposing a fission action once the assembly reaches a maximal size, Nmax, usually in the order of NG. This splitting action produces two progeny of same size, and one of which is grown again.

The model is subjected to a Monte Carlo algorithm based simulations, using Gillespie algorithm.

Selection

In 2010, Eors Szathmary and collaborators have chosen GARD as an archetypal metabolism-first realization. They have introduced selection coefficient into the model, which increase or decrease the growth rate of assemblies, depending on how similar or dis-similar they are to a given target. They found that the ranking of the assemblies are un-affected by the selection pressure, and concluded that GARD does not exhibit Darwinian evolution.

In 2012 it was shown that this criticism is erroneous and was refuted. Two major drawbacks of the 2010 paper were: (1) they have focused on a general assembly and not on a composome or compotype (faithfully replicating and quasispecies, respectively); (2) they have performed only a single, random, simulation to test the selectability.

Quasispecies

The quasispecies model describes a population of replicators that replicate with relatively high mutations. Due to mutations and back mutations the population eventually centres around a master-replicator (master sequence). GARD's populations were shown to form a quasispecies around a master-compotype and to exhibit an Error catastrophe, similarly to classical quasispecis such as RNA viruses .

External links

- GARD10 MATLAB code (see Markovitch and Lancet, 2012): https://github.com/ModelingOriginsofLife/GARD
- Doron Lancet homepage[328] at Weizmann Institute of Science, who is the inventor of GARD.
- Origin of life (OOL[329]) at the Weizmann Institute.

PAH world hypothesis

The **PAH world hypothesis** is a speculative hypothesis that proposes that poly-cyclic aromatic hydrocarbons (PAH), known to be abundant in the universe, including in comets, and, as well, assumed to be abundant in the primordial soup of the early Earth, played a major role in the origin of life by mediating the synthesis of RNA molecules, leading into the RNA world. However, as yet, the hypothesis is untested.

Background

The Miller–Urey experiment in 1952, and others since, demonstrated the syn-thesis of organic compounds, such as amino acids, formaldehyde and sugars, from the original inorganic precursors the researchers presumed to have been present in the primordial soup (but is no longer considered likely). This exper-iment inspired many others. In 1961, Joan Oró found that the nucleotide base adenine could be made from hydrogen cyanide (HCN) and ammonia in a water solution. Experiments conducted later showed that the other RNA and DNA nucleobases could be obtained through simulated prebiotic chemistry with a reducing atmosphere.

The RNA world hypothesis shows how RNA can become its own catalyst (a ribozyme). In between there are some missing steps such as how the first RNA molecules could be formed. The PAH world hypothesis was proposed by Simon Nicholas Platts in May 2004 to try to fill in this missing step. A more thoroughly elaborated idea has been published by Ehrenfreund *et al.*.

Polycyclic aromatic hydrocarbons

Polycyclic aromatic hydrocarbons are the most common and abundant of the known polyatomic molecules in the visible universe, and are considered a likely constituent of the primordial sea. PAHs, along with fullerenes (or "buck-yballs"), have been recently detected in nebulae. (Fullerenes are also impli-cated in the origin of life; according to astronomer Letizia Stanghellini, "It's possible that buckyballs from outer space provided seeds for life on Earth.") In September 2012, NASA scientists reported that PAHs, subjected to interstellar medium (ISM) conditions, are transformed, through hydrogenation, oxygena-tion and hydroxylation, to more complex organics — "a step along the path toward amino acids and nucleotides, the raw materials of proteins and DNA, respectively". Further, as a result of these transformations, the PAHs lose their spectroscopic signature which could be one of the reasons "for the lack of PAH

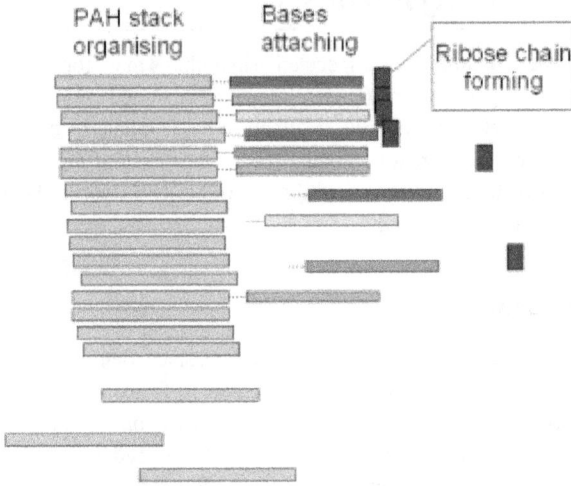

Figure 69: *A PAH stack assembling*

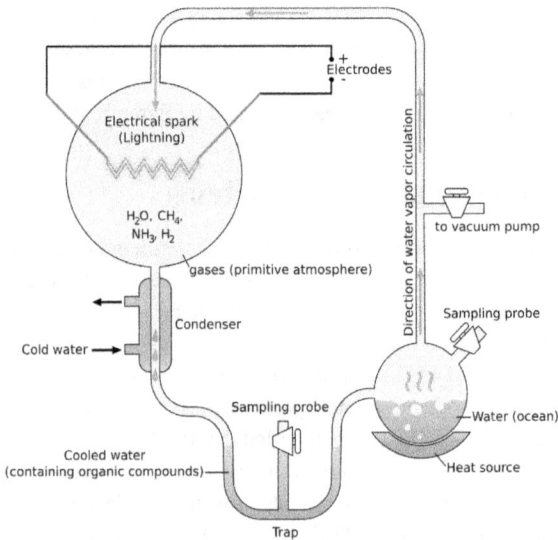

Figure 70: *The Miller–Urey experiment showed that organic compounds can be readily produced under the presumed conditions of the early Earth*

detection in interstellar ice grains, particularly the outer regions of cold, dense clouds or the upper molecular layers of protoplanetary disks."

On June 6, 2013, scientists at the IAA-CSIC reported the detection of polycyclic aromatic hydrocarbons in the upper atmosphere of Titan, the largest moon of the planet Saturn.

PAHs are not normally very soluble in sea water, but when subject to ionizing radiation such as solar UV light, the outer hydrogen atoms can be stripped off and replaced with a hydroxyl group, rendering the PAHs far more soluble in water.

These modified PAHs are amphiphilic, which means that they have parts that are both hydrophilic and hydrophobic. When in solution, they assemble in discotic mesogenic (liquid crystal) stacks which, like lipids, tend to organize with their hydrophobic parts protected.

On February 21, 2014, NASA announced a greatly upgraded database[330] for tracking polycyclic aromatic hydrocarbons (PAHs) in the universe. More than 20% of the carbon in the universe may be associated with PAHs, possible starting materials for the formation of life. PAHs seem to have been formed as early as a couple of billion years after the Big Bang, are abundant in the universe, and are associated with new stars and exoplanets.

Attachment of nucleobases to PAH scaffolding

In the self-ordering PAH stack, the separation between adjacent rings is 0.34 nm. This is the same separation found between adjacent nucleotides of RNA and DNA. Smaller molecules will naturally attach themselves to the PAH rings. However PAH rings, while forming, tend to swivel around on one another, which will tend to dislodge attached compounds that would collide with those attached to those above and below. Therefore, it encourages preferential attachment of flat molecules such as pyrimidine and purine nucleobases, the key constituents (and information carriers) of RNA and DNA. These bases are similarly amphiphilic and so also tend to line up in similar stacks.

Attachment of oligomeric backbone

According to the hypothesis, once the nucleobases are attached (via hydrogen bonds) to the PAH scaffolding, the inter-base distance would select for "linker" molecules of a specific size, such as small formaldehyde (methanal) oligomers, also taken from the prebiotic "soup", which will bind (via covalent bonds) to the nucleobases as well as each other to add a flexible structural backbone.[331,332]

Detachment of the RNA-like strands

A subsequent transient drop in the ambient pH (increase in acidity), for example as a result of a volcanic discharge of acidic gases such as sulfur dioxide or carbon dioxide, would allow the bases to break off from their PAH scaffolding, forming RNA-like molecules (with the formaldehyde backbone instead of the ribose-phosphate backbone used by "modern" RNA, but the same 0.34 nm pitch).

Formation of ribozyme-like structures

The hypothesis further speculates that once long RNA-like single strands are detached from the PAH stacks, and after ambient pH levels became less acidic, they would tend to fold back on themselves, with complementary sequences of nucleobases preferentially seeking out each other and forming hydrogen bonds, creating stable, at least partially double-stranded RNA-like structures, similar to ribozymes. The formaldehyde oligomers would eventually be replaced with more stable ribose-phosphate molecules for the backbone material, resulting in a starting milestone for the RNA world hypothesis, which speculates about further evolutionary developments from that point.

External links

- Life's ingredients found in early universe[333] New Scientist Magazine 14:49 July 29, 2005
- RNA-directed amino acid homochirality[334]

Appendix

References

[1] Compare:

[2] Also occasionally called biopoiesis.

[3] Conference held at League City, TX

[4] //en.wikipedia.org/w/index.php?title=Abiogenesis&action=edit

[5] Ward, Peter (2005), "Life as we do not know it" (Viking Books)

[6] Early edition, published online before print.

[7] M.D> Brasier (2012), "Secret Chambers: The Inside Story of Cells and Complex Life" (Oxford Uni Press), p.298

[8] Ward, Peter & Kirschvink, Joe, op cit, p.42

[9] : "It is often said that all the conditions for the first production of a living organism are now present, which could ever have been present. But if (and oh! what a big if!) we could conceive in some warm little pond, with all sorts of ammonia and phosphoric salts, light, heat, electricity, , present, that a compound was chemically formed ready to undergo still more complex changes, at the present day such matter would be instantly devoured or absorbed, which would not have been the case before living creatures were formed." — Charles Darwin, 1 February 1871

[10] Abiogenesis - Definition https://link.springer.com/referenceworkentry/10.1007/978-3-642-27833-4_2-4. 20 April 2015. *Encyclopedia of Astrobiology.*

[11] Coughlan, Andy (2016) "Smallest ever genome comes to life: Humans built it but we don't know what a third of its genes actually do" (New Scientist 2nd April 2016 No 3067)p.6

[12] Post is reprinted from materials provided by the Rochester Institute of Technology.

[13] Cavitation-Induced Synthesis of Biogenic Molecules on Primordial Earth https://pubs.acs.org/doi/abs/10.1021/acscentsci.7b00325. Natan-Haim Kalson, David Furman, and Yehuda Zeiri. ACS Publications. ACS Cent. Sci., September 11, 2017, 3 (9), pp 1041–1049.

[14] The reactions are:

$$FeS + H_2S \rightarrow FeS_2 + 2H^+ + 2e^-$$
$$FeS + H_2S + CO_2 \rightarrow FeS_2 + HCOOH$$

[15] "Special Issue: Proceedings from the Eighteenth International Symposium on Chirality (ISCD-18), Busan, Korea, 2006"

[16] • "Early View (Online Version of Record published before inclusion in an issue)"

[17]

[18] Nature Vol 535, 28 July 2016,"Early Life Liked it Hot", p.468

[19] For a deeper integrative version of this hypothesis, see in particular , interconnecting the "Two RNA worlds" concept and other detailed aspects; and

[20] The reactions are:

Reaction 1: Fayalite + water → magnetite + aqueous silica + hydrogen
$$3Fe_2SiO_4 + 2H_2O \rightarrow 2Fe_3O_4 + 3SiO_2 + 2H_2$$
Reaction 2: *Forsterite + aqueous silica → serpentine*
$$3Mg_2SiO_4 + SiO_2 + 4H_2O \rightarrow 2Mg_3Si_2O_5(OH)_4$$
Reaction 3: *Forsterite + water → serpentine + brucite*
$$2Mg_2SiO_4 + 3H_2O \rightarrow Mg_3Si_2O_5(OH)_4 + Mg(OH)_2$$
Reaction 3 describes the hydration of olivine with water only to yield serpentine and $Mg(OH)_2$ (brucite). Serpentine is stable at high pH in the presence of brucite like calcium silicate hydrate, (C-S-H) phases formed along with portlandite ($Ca(OH)_2$) in hardened Portland cement paste after the hydration of belite (Ca_2SiO_4), the artificial calcium equivalent of forsterite. Analogy of reaction 3 with belite hydration in ordinary Portland cement: *Belite + water → C-S-H phase + portlandite*
$$2 Ca_2SiO_4 + 4 H_2O \rightarrow 3 CaO \cdot 2 SiO_2 \cdot 3 H_2O + Ca(OH)_2$$

[21] Shock, E. L. & Boyd, E. S. "Geomicrobiology and microbial geochemistry:principles of geobiochemistry. *Elements* 11, 389 –394 (2015).

[22] Nature, Vol 535, 28 July 2016. p.468

[23] Rampelotto, P. H. (2010). Panspermia: A promising field of research. In: Astrobiology Science Conference. Abs 5224.

[24] Forward planetary contamination like *Tersicoccus phoenicis*, that has shown resistance to methods usually used in spacecraft assembly clean rooms:

[25] "Paper presented at the Symposium 'Astrochemistry: molecules in space and time' (Rome, 4–5 November 2010), sponsored by Fondazione 'Guido Donegani', Accademia Nazionale dei Lincei."

[26] Post is reprinted from materials provided by The University of Hong Kong.

[27] Special Issue: "Origin of Life 2011"

[28]

[29] American Astronomical Society Meeting 203, #110.17, January 2004.

[30] Boltzmann, L. (1886) The Second Law of Thermodynamics, in: Ludwig Boltzmann: Theoretical physics and Selected writings, edited by: McGinness, B., D. Reidel, Dordrecht, The Netherlands, 1974.

[31] Schrödinger, Erwin (1944) What is Life? The Physical Aspect of the Living Cell. Cambridge University Press

[32] Onsager, L. (1931) Reciprocal Relations in Irreversible Processes I and II, *Phys. Rev.* 37, 405; 38, 2265 (1931)

[33] Prigogine, I. (1967) An Introduction to the Thermodynamics of Irreversible Processes, Wiley, New York

[34] Unrean, P., Srienc, F. (2011) Metabolic networks evolve towards states of maximum entropy production, Metabolic Engineering 13, 666-673.

[35] Zotin, A. I. (1984) Bioenergetic trends of evolutionary progress of organisms, in: Thermodynamics and regulation of biological processes, edited by: Lamprecht, I. and Zotin, A. I., De Gruyter, Berlin, 451-458.

[36] Michaelian, K. (2016) Thermodynamic Dissipation Theory of the Origin and Evolution of Life: Salient characteristics of RNA and DNA and other fundamental molecules suggest an origin of life driven by UV-C light, Printed by CreateSpace, Mexico City, ,

[37] Sagan, C. (1973) Ultraviolet Selection Pressure on the Earliest Organisms, J. Theor. Biol., 39, 195-200.

[38] Michaelian, K. and Simeonov, A. (2015) Fundamental molecules of life are pigments which arose and evolved to dissipate the solar spectrum. Cornell ArXiv arXiv:1405.4059v2 [physics.bio-ph]

[39] http//download.springer.com

[40] //lccn.loc.gov/2008933212

[41] //lccn.loc.gov/2008030270

[42] //www.worldcat.org/oclc/225874308

[43] //lccn.loc.gov/2007010767

[44] //www.worldcat.org/oclc/86090399

[45] https://archive.org/details/modesoforiginofl00bast

[46] //lccn.loc.gov/11004276

[47] //www.worldcat.org/oclc/42959303

[48] //lccn.loc.gov/51005794

[49] //lccn.loc.gov/60013823

[50] //lccn.loc.gov/67098482

[51] //lccn.loc.gov/96031351

[52] //lccn.loc.gov/98038295

[53] //www.worldcat.org/oclc/55589795

[54] //lccn.loc.gov/70415289

[55] //www.worldcat.org/oclc/25220

[56] //doi.org/10.1007/978-3-642-19598-3_12

[57] //www.worldcat.org/issn/2192-4791

[58] //lccn.loc.gov/2013950482

[59] //www.worldcat.org/oclc/900189038

[60] //lccn.loc.gov/2007013738

[61] //www.worldcat.org/oclc/173522612

[62] http://adsabs.harvard.edu/abs/2001GSLSP.190..205D
[63] //doi.org/10.1144/gsl.sp.2001.190.01.14
[64] //www.worldcat.org/issn/0305-8719
[65] //lccn.loc.gov/2003464816
[66] //www.worldcat.org/oclc/48570033
[67] //www.worldcat.org/oclc/834491774
[68] //doi.org/10.1016/S0166-2635%2807%2915023-4
[69] //lccn.loc.gov/2009525003
[70] //lccn.loc.gov/96229669
[71] //www.worldcat.org/oclc/35648431
[72] //lccn.loc.gov/2004059864
[73] //www.worldcat.org/oclc/56617123
[74] //lccn.loc.gov/60002548
[75] //lccn.loc.gov/99021079
[76] //lccn.loc.gov/79001315
[77] //www.worldcat.org/oclc/4665354
[78] https://archive.org/details/proceedings00inte
[79] //lccn.loc.gov/59012060
[80] //lccn.loc.gov/2005012839
[81] //www.worldcat.org/oclc/60321860
[82] http://aleph0.clarku.edu/huxley/CE8/B-Ab.html
[83] //lccn.loc.gov/70029958
[84] //lccn.loc.gov/91011148
[85] //www.worldcat.org/oclc/23253930
[86] //lccn.loc.gov/94025268
[87] http://www.panspermia.org/oseti.htm
[88] http://www.coseti.org/4273-sch.htm
[89] //doi.org/10.1117/12.435366
[90] //lccn.loc.gov/2001279159
[91] //lccn.loc.gov/2009005046
[92] //www.worldcat.org/oclc/286488326
[93] //doi.org/10.1007/978-3-642-21625-1
[94] //lccn.loc.gov/2011935879
[95] //www.worldcat.org/oclc/733245537
[96] //lccn.loc.gov/00026070
[97] //lccn.loc.gov/98004166
[98] //www.worldcat.org/oclc/38948118
[99] //lccn.loc.gov/53010161
[100] //doi.org/10.1101/SQB.1987.052.01.004
[101] //www.worldcat.org/oclc/19850881
[102] //lccn.loc.gov/2001030052
[103] //www.worldcat.org/oclc/45806501
[104] http://penelope.uchicago.edu/ross/ross210.html
[105] http://www.rbsp.info/rbs/PDF/spie05-telos.pdf
[106] http://spie.org/Publications/Proceedings/Volume/5906?origin_id=x4323&start_year=
2005&end_year=2005
[107] //doi.org/10.1117/12.663480
[108] //lccn.loc.gov/2005284378
[109] https://archive.org/details/fragmenoscien02tyndrich
[110] //www.worldcat.org/oclc/726998155
[111] http://xtf.lib.virginia.edu/xtf/view?docId=DicHist/uvaGenText/tei/DicHist4.xml;chunk.id=
dv4-39;toc.depth=1;toc.id=dv4-39;brand=default
[112] //lccn.loc.gov/72007943
[113] //lccn.loc.gov/2003269978
[114] //lccn.loc.gov/69010418
[115] //lccn.loc.gov/2009044011

[116] http://isotope.colorado.edu/~mojzsis/1997_Arrhenius_JTB.pdf
[117] //doi.org/10.1006/jtbi.1996.0385
[118] //www.worldcat.org/issn/0022-5193
[119] //www.ncbi.nlm.nih.gov/pubmed/9299295
[120] //www.ncbi.nlm.nih.gov/pmc/articles/PMC1578732
[121] //doi.org/10.1098/rstb.2006.1842
[122] //www.worldcat.org/issn/0962-8436
[123] //www.ncbi.nlm.nih.gov/pubmed/16754610
[124] //doi.org/10.1016/j.jtbi.2007.01.028
[125] //www.ncbi.nlm.nih.gov/pubmed/17399743
[126] //doi.org/10.1016/j.cub.2016.12.001
[127] http://adsabs.harvard.edu/abs/1991SciAm.264b.116H
[128] //doi.org/10.1038/scientificamerican0291-116
[129] //www.worldcat.org/issn/0036-8733
[130] http://www.iiste.org/Journals/index.php/JEES/article/view/9903
[131] //www.worldcat.org/issn/2224-3216
[132] //www.ncbi.nlm.nih.gov/pmc/articles/PMC1664691
[133] //doi.org/10.1098/rstb.2006.1909
[134] //www.ncbi.nlm.nih.gov/pubmed/17008225
[135] http://news.discovery.com/earth/oceans/life-pond-ocean-122402.htm
[136] https://web.archive.org/web/20120217040600/http://astrobiology.ucla.edu/pages/res3b.html
[137] http://astrobiology.ucla.edu/pages/res3b.html
[138] https://web.archive.org/web/20071221000659/http://exobio.ucsd.edu/arrhenius.htm
[139] http://exobio.ucsd.edu/arrhenius.htm
[140] http://www.whatislife.com/about.html
[141] //doi.org/10.1002/1522-2675%2820000906%2983%3A9%3C2398%3A%3AAID-HLCA2398%3E3.0.CO%3B2-5
[142] //www.worldcat.org/issn/0018-019X
[143] //www.ncbi.nlm.nih.gov/pubmed/11543578
[144] //www.ncbi.nlm.nih.gov/pmc/articles/PMC3203773
[145] http://adsabs.harvard.edu/abs/2011PNAS..10817639P
[146] //doi.org/10.1073/pnas.1108061108
[147] //www.worldcat.org/issn/0027-8424
[148] //www.ncbi.nlm.nih.gov/pubmed/22006301
[149] http://adsabs.harvard.edu/abs/1988Natur.336..117R
[150] //doi.org/10.1038/336117a0
[151] //www.worldcat.org/issn/0028-0836
[152] http://www.igpp.ucla.edu/public/mkivelso/refs/PUBLICATIONS/shcok%20hiT%20life%20Mars7JE01087.pdf
[153] http://adsabs.harvard.edu/abs/1997JGR...10223687S
[154] //doi.org/10.1029/97je01087
[155] //www.worldcat.org/issn/0148-0227
[156] http://exploringorigins.org/
[157] http://www.gla.ac.uk/projects/originoflife/
[158] http://www.earthfacts.com/evolution-and-life/howlifebeganearth/
[159] https://www.bbc.co.uk/programmes/p004y29f
[160] Four diagrams of evolutionary models
[161] https://cosmosmagazine.com/palaeontology/measuring-sixth-mass-extinction
[162] http://www.bbc.co.uk/nature/history_of_the_earth
[163] https://cosmosmagazine.com/palaeontology/big-five-extinctions
[164] "Oldest traces of life on Earth found in Quebec, dating back roughly 3.8 billion years", by Nicole Mortilanno, CBC News http://www.cbc.ca/news/technology/oldest-record-life-earth-found-quebec-1.4004545
[165] http://www.nature.com/scitable/topicpage/why-are-cells-powered-by-proton-gradients-14373960
[166] //lccn.loc.gov/2007010767

[167] //www.worldcat.org/oclc/86090399

[168] https://www.novapublishers.com/catalog/product_info.php?products_id=31918

[169] //lccn.loc.gov/2011038504

[170] //www.worldcat.org/oclc/828424701

[171] //lccn.loc.gov/2004022738

[172] //www.worldcat.org/oclc/56672295

[173] http://web.gps.caltech.edu/~jkirschvink/pdfs/firstsnowball.pdf

[174] //lccn.loc.gov/91015085

[175] //www.worldcat.org/oclc/23583672

[176] //lccn.loc.gov/96071014

[177] //www.worldcat.org/oclc/36442106

[178] //lccn.loc.gov/2011934330

[179] //www.worldcat.org/oclc/741539226

[180] //lccn.loc.gov/98034087

[181] //www.worldcat.org/oclc/47011675

[182] //lccn.loc.gov/2004059864

[183] //www.worldcat.org/oclc/56617123

[184] http://evolution.berkeley.edu/

[185] http://tolweb.org/Life_on_Earth/1

[186] http://www.talkorigins.org/origins/geo_timeline.html

[187] http://www.palaeos.com

[188] http://www.johnkyrk.com/evolution.html

[189] http://sci.waikato.ac.nz/evolution/plantEvolution.shtml

[190] http://sci.waikato.ac.nz/evolution/AnimalEvolution.shtml

[191] https://web.archive.org/web/20150315040156/http://draget.net/hoe/index.php

[192] http://draget.net/hoe/index.php

[193] http://exploringtime.org/?page=segments

[194] http://www.plantsci.cam.ac.uk/timeline/

[195] http://www.naturelyrics.com/pages/articles/nature_photography/nature_in_nature_photography.html

[196] https://web.archive.org/web/20140715055239/http://taxonomicon.taxonomy.nl/TaxonTree.aspx?id=1&src=0

[197] https://web.archive.org/web/20051222163318/http://sn2000.taxonomy.nl/Main/Classification/1.htm

[198] https://www.biolib.cz/en/taxon/id14772

[199] https://www.google.com/search?q=earliest+known+life+forms&source=lnms&tbm=isch&sa=X&ved=0ahUKEwjR9YXu36HbAhXyuFkKHXUFAmkQ_AUICigB&biw=1800&bih=961

[200] Tyndall, John; Fragments of Science, Vol 2, chapters IV, XII (1876), XIII(1878); Pub. P. F. Collier, New York 1905; (Available at: https://archive.org/details/fragmenoscien02tyndrich)

[201] Zubay, Geoffrey. Origins of Life, Second Edition: On Earth and in the Cosmos. Academic Press 2000.

[202] Levine, Russell and Evers, Chris (1999). "The Slow Death of Spontaneous Generation" http://webprojects.oit.ncsu.edu/project/bio183de/Black/cellintro/cellintro_reading/Spontaneous_Generation.html.The National Health Museum.

[203] Stillingfleet, Edward. Origines Sacrae. Cambridge University Press 1697. May be downloaded from https://archive.org/details/originessacraeor00stil

[204] Guthrie, William Keith Chambers, The Presocratic Tradition from Parmenides to Democritus, p. 13, .

[205] Seyffert, Oskar (1894), Dictionary of Classical Antiquities, page 480

[206] Censorinus, De Die Natali, IV, 7

[207] http://internetshakespeare.uvic.ca/Library/Texts/Ant/F1/Scene/2.7

[208] Royal Society (Great Britain); Hutton, Charles, 1737-1823; Shaw, George, 1751-1813; Pearson, Richard, 1765-1836. The Extract of a Letter written by Mr. JOHN RAY, to the Editor, from Middleton, July 3, 1671, concerning Spontaneous Generation;... Number 73,

p. 2219. The Philosophical transactions of the Royal Society of London, from their commencement in 1665, in the year 1800. Page 617-618. May be downloaded from: https://archive.org/details/philosophicaltra01royarich

[209] Priestley, Joseph. Observations and Experiments relating to equivocal, or spontaneous, Generation. Transactions of the American Philosophical Society, Volume VI, page 119-129, 1809. Download from: https://archive.org/details/transactionsofam61809amer

[210] Rennie, James. Insect Transformations. Page 10. Pub: Charles Knight 1838 Download from: https://archive.org/details/insecttransforma00renn

[211] Pasteur's Papers on the Germ Theory http://biotech.law.lsu.edu/cphl/history/articles/pasteur.htm

[212] Louis Pasteur: External links

[213] The controversy over spontaneous generation http://www.studentsguide.in/microbiology/microbiology/controversy-over-spontaneous-generation.html

[214] The microbial world: a look at things small http://www.microbiologytext.com/index.php?module=Book&func=displayarticle&art_id=27

[215] Biogenesis and Abiogenesis: Critiques and Addresses http://aleph0.clarku.edu/huxley/CE8/B-Ab.html

[216] Experiments on origin of organic molecules http://nitro.biosci.arizona.edu/courses/EEB105/lectures/Origins_of_Life/origins.html . Retrieved 13 January 2008.

[217] Proteinoid microspheres and methods for preparation and use thereof https://www.google.com/patents/US5601846. Patent US 5601846 A (1997)

[218] http://adsabs.harvard.edu/abs/1958Sci...128.1214F

[219] //doi.org/10.1126/science.128.3333.1214

[220] //www.jstor.org/stable/1756313

[221] Steinfeld J.I., Francisco J.S. and Hase W.L. *Chemical Kinetics and Dynamics* (2nd ed., Prentice-Hall 1999) p.151-2

[222] Moore J.W. and Pearson R.G. *Kinetics and Mechanism* (John Wiley 1981) p.26

[223] Espenson, J.H. *Chemical Kinetics and Reaction Mechanisms* (2nd ed., McGraw-Hill 2002) p.190

[224] Dynamics of the Brusselator

[225] Ecology, the Ascendent Perspective", Robert Ulanowicz, Columbia Univ. Press 1997.

[226] Investigations, Stuart Kauffman.

[227] http://www.eeng.dcu.ie/~alife/bmcm9901/html-multi/

[228] //arxiv.org/abs/adap-org/9809003

[229] http://adsabs.harvard.edu/abs/1998PhRvL..81.5684J

[230] //doi.org/10.1103/PhysRevLett.81.5684

[231] "Special Issue: Proceedings from the Eighteenth International Symposium on Chirality (ISCD-18), Busan, Korea, 2006"

[232] However, the message can be confusing. In , the entry for *Enantiomerically Pure/Enantiopure* says "Use of homochiral as a synonym is strongly discouraged"; but the entry for *Homochiral* says "See *enantiomerically pure/enantiopure*."

[233] https://web.archive.org/web/20101130094258/http://scienceweek.com/1998/sw980828.htm

[234] http://scienceweek.com/1998/sw980828.htm

[235] https://www.imperial.ac.uk/college.asp?P=5343

[236] //www.jstor.org/stable/24996649

[237] http://adsabs.harvard.edu/abs/2017PhRvE..95c2407J

[238] //doi.org/10.1103/PhysRevE.95.032407

[239] http://pubs.acs.org/cen/news/8224/8225earlyscicon.html

[240] https://web.archive.org/web/20041222030845/http://www.photonics.com/spectra/tech/XQ/ASP/techid.443/QX/read.htm

[241] http://agenda.albanova.se/conferenceDisplay.py?confId=322

[242] //en.wikipedia.org/w/index.php?title=Template:Evolutionary_biology&action=edit

[243] Deamer, D.W.; Dworkin, J.P. "Chemistry and Physics of Primitive Membranes". Top. Curr. Chem. 2005, 259, 1–27.

[244] Walde, P. "Surfactant Assemblies and their various possible roles for the origin(s) of life." Orig. Life Evol. Biosph. 2006, 36, 109–150.

[245] http://www.biofaction.com/synth-ethic/?p=61

[246] Woese C.R. (1967). The genetic code: The molecular basis for genetic expression. p. 186. Harper & Row

[247] Hani S. Zaher and Peter J. Unrau, Selection of an improved RNA polymerase ribozyme with superior extension and fidelity. http://www.rnajournal.org/cgi/content/abstract/13/7/1017 RNA (2007), 13:1017-1026

[248] Patton, John T. Editor (2008). Segmented Double-stranded RNA Viruses: Structure and Molecular Biology. Caister Academic Press. Editor's affiliation: Laboratory of Infectious Diseases, NIAID, NIH, Bethesda, MD 20892-8026.

[249] Bell, Graham: The Basics of Selection. Springer, 1997.

[250] Carole Anastasi, Michael A. Crowe, Matthew W. Powner, John D. Sutherland "Direct Assembly of Nucleoside Precursors from Two- and Three-Carbon Units *Angewandte Chemie International Edition* **45**(37):6176–79, 2006.

[251] Holmes, Bob (2012) "First Glimpse at the birth of DNA" (New Scientist April 12, 2012)

[252] Woese CR (1983). The primary lines of descent and the universal ancestor. Chapter in pp. 209-233.

[253] see pgs. 293-297

[254] Platts, Simon Nicholas, "The PAH World – Discotic polynuclear aromatic compounds as a mesophase scaffolding at the origin of life" http://www.pahworld.com/

[255] Allamandola, Louis et Al. "Cosmic Distribution of Chemical Complexity" http://amesteam.arc.nasa.gov/Research/cosmic.html

[256] "Phosphorylation, Oligomerization and Self-assembly in Water Under Potential Prebiotic Conditions", Gibard et al., Nature Chemistry (2017) doi:10.1038/nchem.2878, published online 06 November 2017 https://www.nature.com/articles/nchem.2878

[257] //doi.org/10.7554/eLife.35255

[258] //www.ncbi.nlm.nih.gov/pubmed/29759114

[259] http://adsabs.harvard.edu/abs/1994SciAm.271d..76O

[260] //doi.org/10.1038/scientificamerican1094-76

[261] //www.ncbi.nlm.nih.gov/pubmed/7524147

[262] //doi.org/10.1080/10409230490460765

[263] //www.ncbi.nlm.nih.gov/pubmed/15217990

[264] http://adsabs.harvard.edu/abs/2005JMolE..61..264V

[265] //doi.org/10.1007/s00239-004-0362-7

[266] //www.ncbi.nlm.nih.gov/pubmed/16044244

[267] http://cshperspectives.cshlp.org/content/2/12/a002196.full.pdf

[268] //doi.org/10.1101/cshperspect.a002196

[269] //www.ncbi.nlm.nih.gov/pmc/articles/PMC2982173

[270] //www.ncbi.nlm.nih.gov/pubmed/20462999

[271] http://www.biology-direct.com/content/7/1/23

[272] //doi.org/10.1186/1745-6150-7-23

[273] //www.ncbi.nlm.nih.gov/pmc/articles/PMC3495036

[274] //www.ncbi.nlm.nih.gov/pubmed/22793875

[275] //www.ncbi.nlm.nih.gov/pmc/articles/PMC2845210

[276] //doi.org/10.1101/cshperspect.a005439

[277] //www.ncbi.nlm.nih.gov/pubmed/20452951

[278] http://exploringorigins.org/rnaworld.html

[279] https://web.archive.org/web/20120301174206/http://www.origins.rpi.edu/chem.html

[280] http://www.origins.rpi.edu/chem.html

[281] https://www.nobelprize.org/nobel_prizes/chemistry/laureates/1989/altman-article.html

[282] http://www.hhmi.org/sites/default/files/Bulletin/2002/June/june2002_fulltext.pdf

[283] https://web.archive.org/web/20040522184700/http://www.hhmi.org/bulletin/pdf/june2002/RNA.pdf

[284] https://www.nobelprize.org/nobel_prizes/chemistry/laureates/1989/cech-article.html

[285] (requires nonfree AAAS member subscription)

[286] Russell, Michael (Ed), (2010), "Origins, Abiogenesis and the Search for Life in the Universe" (Cosmology Science Publications)

[287] Forward planetary contamination like *Tersicoccus phoenicis*, that has shown resistance to methods usually used in spacecraft assembly clean rooms:

[288] A variation of the panspermia hypothesis is **necropanspermia** which astronomer Paul Wesson describes as follows: "The vast majority of organisms reach a new home in the Milky Way in a technically dead state ... Resurrection may, however, be possible."

[289] Hoyle, F. and Wickramasinghe, N.C. (1981). *Evolution from Space*. Simon & Schuster Inc., NY, and J.M. Dent and Son, London (1981), ch3 pp. 35–49.

[290] Wickramasinghe, J., Wickramasinghe, C. and Napier, W. (2010). *Comets and the Origin of Life* http://alpha.sinp.msu.ru/~panov/LibBooks/LIFE/10972__legalreads.com.pdf. World Scientific, Singapore. ch. 6 pp. 137–154.

[291] Margaret O'Leary (2008) Anaxagoras and the Origin of Panspermia Theory, iUniverse publishing Group,

[292] Arrhenius, S. (1908) *Worlds in the Making: The Evolution of the Universe*. New York, Harper & Row.

[293] Early edition, published online before print.

[294] Studies Focus On Spacecraft Sterilization https://web.archive.org/web/20060502194219/http://www.aero.org/news/newsitems/sterilization073001.html. aero.org (July 30, 2000)

[295] Dry heat sterilisation process to high temperatures https://web.archive.org/web/20120201224127/http://www.esa.int/esaMI/Aurora/SEMBJG9ATME_0.html. European Space Agency (22 May 2006)

[296] Slow-moving rocks better odds that life crashed to Earth from space http://www.princeton.edu/main/news/archive/S34/82/42M30/. News at Princeton, September 24, 2012.

[297] "Die Verbreitung des Lebens im Weltenraum" (the "Distribution of Life in Space"). Published in *Die Umschau*. 1903.

[298] Gold, T. "Cosmic Garbage", Air Force and Space Digest, 65 (May 1960).

[299] " Anticipating an RNA world. Some past speculations on the origin of life: where are they today? http://www.fasebj.org/cgi/reprint/7/1/238.pdf" by L. E. Orgel and F. H. C. Crick in *FASEB J.* (1993) Volume 7 pages 238–239.

[300] Finley, Dave (February 28, 2013) Discoveries Suggest Icy Cosmic Start for Amino Acids and DNA Ingredients https://www.nrao.edu/pr/2013/newchem/. *The National Radio Astronomy Observatory*

[301] http://www.astrochem.org/pahdb/

[302] http://sci.esa.int/rosetta/57863-altwegg-et-al-2016/

[303] Discovery of New Microorganisms in the Stratosphere http://www.physorg.com/news156626262.html. Physorg (March 18, 2009)

[304] *A.A. Imshenetsky, S.V. Lysenko, G.A. Kazakov, "Upper boundary of the biosphere," Appl Environ Microbiol, vol. 35, pp. 1-5, 1978. • A.A. Imshenetsky, S.V. Lysenko, G.A. Kazakov, N.V. Ramkova, "On micro-organisms of the stratosphere," Life Sci Space Res, vol. 14, pp. 359–362, 1976. • Y. Yang, T. Itoh, S. Yokobori, et al., "Deinococcus aetherius sp. nov., isolated from the stratosphere," Int J Syst Evol Microbiol, vol. 60, pp. 776–779, 2010. • S. Shivaji, S. Ara, S.K. Singh, et al., "Draft genome sequence of Bacillus isronensis strain B3W22, isolated from the upper atmosphere," J Bacteriol, vol. 194, pp. 6624–6625, 2012.

[305]

[306] Surviving the Final Frontier http://www.astrobio.net/exclusive/318/surviving-the-final-frontier. astrobio.net (25 November 2002).

[307] Bacterium revived from 25 million year sleep http://commtechlab.msu.edu/sites/dlc-me/news/ns595ap1.html Digital Center for Microbial Ecology

[308] Yokobori, Shin-ichi et al (2010) Microbe space exposure experiment at International Space Station (ISS) proposed in "Tanpopo" mission https://www.researchgate.net/publication/241270775_Microbe_space_exposure_experiment_at_International_Space_Station_(ISS)_proposed_in_Tanpopo_mission. Research Gate.

[309] Yano, H. et al. (2014) " Tanpopo Experiment for Wastrobiology Exposure and Micrometeoroid Capture Onboard the ISS-JEM Exposed Facility http://www.hou.usra.edu/meetings/lpsc2014/pdf/2934.pdf." 45th Lunar and Planetary Science Conference.

[310] Tanpopo mission to search space for origins of life http://the-japan-news.com/news/article/0002066967. *The Japan News*, April 16, 2015.

[311] http://web.snauka.ru/en/issues/2013/12/30018

[312] http://profiles.nlm.nih.gov/SC/B/B/Y/P/_/scbbyp.pdf

[313] //doi.org/10.1038/news040216-20

[314] http://www.scientificamerican.com/article.cfm?id=did-life-come-from-anothe

[315] http://adsabs.harvard.edu/abs/2005SciAm.293e..64W

[316] //doi.org/10.1038/scientificamerican1105-64

[317] N.C. Wickramasinghe, Formaldehyde Polymers in Interstellar Space, Nature, 252, 462, 1974

[318] F. Hoyle and N.C. Wickramasinghe, Identification of the lambda 2200Å interstellar absorption feature, Nature, 270, 323, 1977

[319] Interstellar chemistry makes use of quantum shortcut | Science News http://www.sciencenews. org/view/generic/id/351444/description/Interstellar_chemistry_makes_use_of_quantum_ shortcut#comment_351468

[320] NH4+ in the ISM http://www.astro.uni-koeln.de/site/vorhersagen/molecules/ism/Ammonium. html

[321] Detection Of The Ammonium Ion In Space - Iopscience http://iopscience.iop.org/2041-8205/ 771/1/L10/

[322] First Detection of Methyl Alcohol in a Planet-forming Disc http://astrobiology.com/2016/06/ first-detection-of-methyl-alcohol-in-a-planet-forming-disc.html. 15 June 2016.

[323] http://www.astrochymist.org/astrochymist_ism.html

[324] http://www.astro.uni-koeln.de/cdms/molecules/

[325] http://science.gsfc.nasa.gov/691/cosmicice/interstellar.html

[326] http://www.cv.nrao.edu/~awootten/allmols.html

[327] http://www.physics.nist.gov/PhysRefData/Micro/Html/contents.html

[328] http://www.weizmann.ac.il/molgen/members/lancet.html

[329] http://ool.weizmann.ac.il

[330] http://www.astrochem.org/pahdb/

[331]

[332]

[333] https://web.archive.org/web/20061019104204/http://www.newscientistspace.com/article.ns? id=dn7754

[334] http://www.fasebj.org/cgi/content/full/12/6/503#F1

Article Sources and Contributors

The sources listed for each article provide more detailed licensing information including the copyright status, the copyright owner, and the license conditions.

Abiogenesis *Source:* https://en.wikipedia.org/w/index.php?oldid=853771590 *License:* Creative Commons Attribution-Share Alike 3.0 *Contributors:* 564dude, A2soup, ANDECA, ARTISAN, ATPhosphate, Abductive, Adûnâi, Alaney2k, Alsosaid1987, Amaurea, Another Believer, Apokryltaros, Aquarius-1, Arjayay, Axl, Azcolvin429, BatteryIncluded, Bear-rings, Black Kite, Ceosad, Charlesdrakew, Cheeseskates, Chhandama, ChiBeta, Chiswick Chap, Chris Capoccia, Ciphers, ClueBot NG, ComicsAreJustAllRight, DMacks, Dan100, Danhash, DennisPietras, Dfcorrea00, Doug Weller, Dr. Sroy, Drbogdan, Eddy Tor, Editor2020, El C, Fama Clamosa, Fmadd, GeoWriter, Gilliam, GoingBatty, Graeme Bartlett, GünniX, HaeB, Harizotoh9, Headbomb, Hoffmannrungethailand, Home Lander, Hongqilim, Huntster, Iacobus, IdreamofJeanie, Ira Leviton, Iridescent, Ita140188, J 1982, Jim1138, John D. Croft, Juditsponer, Just plain Bill, KAP03, Kintetsubuffalo, Kku, Kolbasz, Koza1983, Ksarasofi, LiamEdwards4th, Lx 121, Lydianh, Magioladitis, Magyar25, Maltrópa, McSly, Me, Myself, and I are Here, Mikenorton, Mitch Ames, Murakamifan93, Nanotrix, Narky Blert, Natural RX, Nergaal, Nor Setab, OAnick, Oldstone James, Omegatron, Omermar, Omnipaedista, PaleoNeonate, Paolo.dL, Pure Statistics, Quebec99, Rjwilmsi, Robynthehode, RockMagnetist (DCO visiting scholar), Rowan Forest, Roxy the dog, Scott McNay, Sgconlaw, Signedzzz, Silverstair, Slightsmile, Smeagol 17, Sprlzrd, SucreRouge, Sunrise, Taester08, Tgeorgescu, Tiptopper, Tom.Reding, TwoTwoHello, Tyrol5, UNOwenNYC, Ugly Ketchup, Ugog Nizdast, Uvtarex, Vincent Andres, Vsmith, Wahrmund, White whirlwind, Wikid77, Wikimpan, X1\, Yannickvallee, Zedshort, 78 anonymous edits 1

Timeline of the evolutionary history of life *Source:* https://en.wikipedia.org/w/index.php?oldid=851883596 *License:* Creative Commons Attribution-Share Alike 3.0 *Contributors:* 123ert123ert, 7&6=thirteen, A.amitkumar, Abyssal, Alan G. Archer, Alaney2k, Alton.arts, Anagonzalezzz, Antrocent, Audaciter, Azcolvin429, BatteryIncluded, Bear-rings, Bender235, Beyond My Ken, Bgwhite, CaGlwwWEDymzc7KBQC8u, Charlesdrakew, Chaya5260, Chiswick Chap, Cirt, Clpo13, ClueBot NG, Comp.arch, Danno349, Davemck, David Biddulph, David.moreno72, Dewritech, Discospinster, Dodi 8238, Donner60, Drbogdan, Edvaldig, Emperors Of Random, Excirial, Fanx, Ghostsarememories, Gilliam, Goldratte, Golmschenk, Harizotoh9, I dream of horses, InedibleHulk, Iryna Harpy, J 1982, J947, JaconaFrere, Jelle Gouw, Joe2719, John of Reading, Jonpienaar, Kendrick7, KylieTastic, LizzieBabes419, Lukas2021, MRD2014, Mama meta modal, Marek69, MelbourneStar, Mindmatrix, MusikAnimal, Mz7, OAnick, Oshwah, PhilKnight, Philip Trueman, Pilows2, Pinethicket, Pjvandehaar, PlyrStar93, Raquel Baranow, Renée knows it all, RichardWeiss, Ritoxavi, Rjwilmsi, Rominandreu, Rrburke, Ru4true, Ryan Shakiba, Sarr Cat, SchreiberBike, Sein tao, Shellwood, Skoskav∼enwiki, Sunrise, TAnthony, Theroadislong, Tom.Reding, Tonyphill777, Trappist the monk, Uhregqyrwfhpreuqjwbgnv 2l4tuehjngroivwje, Vsmith, W mac8, WMartin74, Wiae, Wikigratia, WikipediaWriter101, Wjfox2005, Yam ileth, Zellfaze, Zortwort, 162 anonymous edits 51

Earliest known life forms *Source:* https://en.wikipedia.org/w/index.php?oldid=851496935 *License:* Creative Commons Attribution-Share Alike 3.0 *Contributors:* Animalparty, Aquarius-1, Aspening, Ayuta Tonomura, CIAVermont, Chiswick Chap, Darsie42, Davemck, Dpleibovitz, Drbogdan, Graeme Bartlett, Habil zare, Harizotoh9, Holothurion, John Abbe, Kintetsubuffalo, Nwbeeson, OAnick, Robshort, Spike Wilbury, TheSandDoctor, Timpo, Turtlesandfoxes, Ugly Ketchup, Volcanoguy, Whoop whoop pull up, Ysangkok, 8 anonymous edits 73

Spontaneous generation *Source:* https://en.wikipedia.org/w/index.php?oldid=853440149 *License:* Creative Commons Attribution-Share Alike 3.0 *Contributors:* -lolololololololololoolol, A. Parrot, Apokryltaros, Atwarwiththem, Beefyt, Bellerops, Belovedfreak, Bender235, Bennv3771, BlakeKbelt, Bluerasberry, Caballero1967, Carlon, Catgut, Charles Matthews, Chiswick Chap, ClueBot NG, Crystallizedcarbon, Cyberbot II, DGG, Darker Dreams, Dcirovic, Deathawk, Denisarona, Dictonary1, Discospinster, Doug Weller, Download, Drbogdan, Dreth, E3cubestore, Ed Poor, Egsan Bacon, EpicEnder, Eric0928, Excirial, Eyesnore, Fightindaman, Fæ, Gary, GeneralizationsAreBad, GorillaWarfare, Hans Dunkelberg, Harizotoh9, I dream of horses, Intelligentsium, Jackalojohn, JaconaFrere, Jade Green Eyes, James343e, Jamesx12345, Jeffrd10, Jinx69, JonRichfield, Junior5a, Keahapana, King of Hearts, Leszek Jańczuk, MB, Marek69, Mary Mark Ockerbloom, Materialscientist, Midas02, Minaker, Mogism, Mzaki, Novangelis, Oleg-ch, Omeganian, Opencooper, Oshwah, Pdcook, PedR, Phdpeter, Philip Trueman, Ramos1990, Rjwilmsi, SchreiberBike, Scientizzle, SensoryRhyme, Serols, SkepticalRaptor, Solomonfromfinland, Sophie means wisdom, Spontaneoustaco, Staszek Lem, Steinbach, Sunrise, Tbhotch, Technopat, The Anonymouse, Thebiologyprimer, Thecheesykid, Tide rolls, Tmol42, TomS TDotO, Toughpigs, Trusilver, Tstrobaugh, Unchartered, Vieque, Widr, Wikipelli, Wolfymoza, Wykypydya, Zaki1996, Zanoni, أسامة الساعدي, 194 anonymous edits 81

Biogenesis *Source:* https://en.wikipedia.org/w/index.php?oldid=840512332 *License:* Creative Commons Attribution-Share Alike 3.0 *Contributors:* ABF, Ajsherb, Alan Liefting, Alx5000∼enwiki, Anbu121, Apokryltaros, Arjayay, Aunt Entropy, BatteryIncluded, Benbest, Bender235, BevinBrett, Bhlara82, BlueBen1111, Blurpeace, Bmf97, CanadianLinuxUser, Cannolis, Charlesdrakew, Chemistryfan, ClueBot NG, Collinsprojects, DARTH SIDIOUS 2, Denispir, DerHexer, Doug Weller, Dougofborg, Drbogdan, Exemplo347, FergusM1970, Froglich, Fsrusc, GarretKadeDupre, Gbrown2013, Gilliam, GoingBatty, GoldenRing, Grafen, Gralgrathor, Guillaume2303, Hakan Kayı, Headbomb, Heyarnoldman, Hmains, Icairns, IntelligentDesign!12, Ixfd64, JSmith60, JWNoctis, Ja 62, Jinx69, Jojhutton, Jprg1966, Jtisabully, Kwiztas, LilyKitty, MaineMichael, Mcoupal, Me, Myself, and I are Here, Mild Bill Hiccup, Nase, Northern Muriqui, Novangelis, O Danilo, Orfen, Palosirkka, Para, Performedge, Pgburgess, Pinethicket, Pipetricker, Prefix-NA, Qzd, RA0808, Rich Farmbrough, Rjwilmsi, SSolheim, ScottyBerg, SensoryRhyme, Serols, Sophie means wisdom, SpaceFlight89, Staszek Lem, Sunrise, Svick, Templerun, Tempodivalse, TheDevilYouKnow, Thebiologyprimer, TomS TDotO, Tow, Trusilver, Tstrobaugh, Vertium, Vsmith, Welsh-girl-Lowri, What makes a man turn neutral?, WikHead, Woodshed, XP1, YK102, Yachtsman1, Yhju, ماني, ಮಮಕ, 197 anonymous edits 93

Primordial soup *Source:* https://en.wikipedia.org/w/index.php?oldid=847524939 *License:* Creative Commons Attribution-Share Alike 3.0 *Contributors:* 72, Amorymeltzer, Appl3pi3888888, Arjayay, BD2412, Bellerops, Biem, Biwhite2, Chhandama, ClueBot NG, CogitoErgoSum14, DaWarMage, Dcirovic, Dcs002, Decumanus, Discospinster, Donner60, Ernsts, Hrafn, I dream of horses, Jandalhandler, Jorfer, Julesd, Klilidiplomus, LilHelpa, LindsayH, Magioladitis, Michaelwuzthere, P.T.Đ, PhageRules1, Samwise granger, Sandos, Serge020, Sunrise, Vsmith, 39 anonymous edits 96

Proteinoid *Source:* https://en.wikipedia.org/w/index.php?oldid=845182116 *License:* Creative Commons Attribution-Share Alike 3.0 *Contributors:* AManWithNoPlan, Aozf05, BatteryIncluded, Bgwhite, Bryan Derksen, Chrumps, Conversion script, Download, Emre D., Eroica, Evolution and evolvability, Firsfron, Hakan Kayı, Jacopo Werther, Jesse V., Jrosenau, Koza1983, Lankenau, Rjwilmsi, Sensonet, TAnthony, Takowl, Trappist the monk, WaiteDavid137, Welsh, Wiae, Yecril, 17 anonymous edits 100

Autocatalysis *Source:* https://en.wikipedia.org/w/index.php?oldid=841536094 *License:* Creative Commons Attribution-Share Alike 3.0 *Contributors:* Alpha3031, Alékos Elefthérios, Azra99, Bollyjeff, Butonic∼enwiki, Cedar101, Ceyockey, Christian75, Citation bot 1, CommonsDelinker, Complexica, D3, Daniele.tampieri, Dcirovic, DerryTaylor, Dino, Dirac66, Ebaskerv, Ebradsha, EconoPhysicist, Eno-ja, Fuortu, Imsobored, Jamesx12345, Jason Quinn, John Riemann Soong, Jü, Kaiserkarl13, Kernsters, Khazar2, Kku, Knights who say ni, L293D, Lexor, Liamstone, LilHelpa, Limhes, Mion, Misternuvistor, Nathanielvirgo, Olin, Ordermaven, Random832, Raven in Orbit, RedAndr, RedWolf, Rifleman 82, Rjwilmsi, RockMagnetist, STBotD, Salix alba, Scope creep, Shyamal, Skylord a52, Sunrise, Surya Prakash.S.A., Tarcieri, TechnoFaye, TedPavlic, Thumperward, User A1, UtherSRG, Xenoglossophobe, Yahadzija, Yakudza, 佛祖西来 , 48 anonymous edits 103

Homochirality *Source:* https://en.wikipedia.org/w/index.php?oldid=852331835 *License:* Creative Commons Attribution-Share Alike 3.0 *Contributors:* Artoannila, BatteryIncluded, Blackbombchu, Cedar101, Chris the speller, Christian75, Citation bot 1, Craxyxarc, Crito10, DMacks, DOSGuy, Daniel Newby, Dartvox78, Dcirovic, Die Antworten, Dmbius, Drbogdan, HPBiochemie, Hakan Kayı, Headbomb, I am One of Many, Ian Pitchford, JPBoyd, Jengod, Jingxin, Karl 334, Kortoso, Krasss, LabFox, Lesterama, M stone, MadMax, Meaningless, Mikespedia, Mnefliu, NFHC, Nave.notnilc, Northfox, Ohconfucius, Project Osprey, RIT RAJARSHI, Rekleov, Rifleman 82, Rjwilmsi, RockMagnetist (DCO visiting scholar), Rzepa, Saehrimnir, Shosukekojo∼enwiki, Sminthopsis84, Speciate, Sprlzrd, Srednuas Lenoroc, Su-no-G, Sunrise, V8rik, WVhybrid, Wenli, Wickey-nl, Wiki007wiki, 38 anonymous edits 113

Protocell *Source:* https://en.wikipedia.org/w/index.php?oldid=847526798 *License:* Creative Commons Attribution-Share Alike 3.0 *Contributors:* BatteryIncluded, CRGreathouse, Callum radiator, Chiswick Chap, CuriousMind01, Drbogdan, Gabriel syme, IceKarma, Jandalhandler, Kalidasa 777, Me, Myself, and I are Here, NawlinWiki, Redrose64, Rjwilmsi, Rowan Forest, SallyForth123, Vsmith, Χρυσάνθη Λυκούση, 23 anonymous edits 121

RNA world *Source:* https://en.wikipedia.org/w/index.php?oldid=847843219 *License:* Creative Commons Attribution-Share Alike 3.0 *Contributors:* 0xF8E8, 3LizardsinaTrenchcoat, AManWithNoPlan, Aard one, Aetheling, Agesworth, Agricolae, AioftheStorm, Alaney2k, Anrnusna, Arch dude, Ashaels, AstroBio Ben, Audaciter, Audriusa, Avoided, BatteryIncluded, Ben Moore, Bernstein0275, Billhpike, Boghog, Boppet, Brandmeister, CCevol2015, Ceyockey, Chaya5260, Chhandama, Chiswick Chap, ChrisGualtieri, ClueBot NG, ComicsAreJustAllRight, Comp.arch, Dcirovic, Deb, Dialectric, Dienerto, DragonflySixtyseven, Dratman, Drbogdan, Edward321, Erikamit, Ernsts, Ewen, Glane23, Glevum, Graham Beards, Gsandi, HPBiochemie, Haplochromis, Headbomb, HighFlyingFish, Holden15, Jarble, Jdaloner, Jhbdel, John D. Croft, JohnArmagh, Jonesey95, Jruderman, JustinClarkCasey, KAP03, Kalidasa 777, Kiwi128, Kuru, LilHelpa, Lor, Lottamiata, Magioladitis, Mandarax, MartinZ, Me, Myself, and I are Here, Middle 8, Mimihitam, Modest Genius, Mpvdm, Niceguyedc, Nn9888, Notreallydavid, NottNott, OperativeKlause, Paleolithicus, Peggy hopper, Peteruetz, Phacelias, Quintar360, Ringbang, Rjwilmsi, Rozi Sh, Sfan00 IMG, Smartse, Speciate, Spr3sso, Stesmo, Strauss12, Sunrise, SzostakJack, Theemathas, TippyGoomba, Tom.Reding, Trappist the monk, Viriditas, Vsmith, Wetman, 98 anonymous edits 131

Iron–sulfur world hypothesis *Source:* https://en.wikipedia.org/w/index.php?oldid=847266014 *License:* Creative Commons Attribution-Share Alike 3.0 *Contributors:* 84user, AC+79 3888, AManWithNoPlan, Andrew c, AndrewHowse, Art LaPella, AxelBoldt, Barbara Shack, BatteryIncluded, Bender235, Cacycle, Citation bot 1, Citation bot 4, ClassicSC, ClueBot NG, Daniel.Cardenas, Dauto, Donarreiskoffer, Dr.Buttons, Drbogdan, Dthomsen8, EagerToddler39, EvenGreenerFish, Extremophile, Farid320, Favonian, Feline1, FghIJklm, Geekdiva, General Epitaph, GregorB, Grock2, Gustavocarra, Headbomb, Heurgh, IceKarma, Insorak, Iridescent, Jasonanaggie, John D. Croft, Julianonions, Koavf, Lexor, Linakieper, Lumos3, MarkGT, MegaHasher,

Image Sources, Licenses and Contributors

The sources listed for each image provide more detailed licensing information including the copyright status, the copyright owner, and the license conditions.

Figure 1 *Source:* https://en.wikipedia.org/w/index.php?title=File:Champagne_vent_white_smokers.jpg *License:* Public Domain *Contributors:* NOAA ... 2
Image *Source:* https://en.wikipedia.org/w/index.php?title=File:Interactive_icon.svg *Contributors:* User:Evolution and evolvability ... 4
Figure 2 *Source:* https://en.wikipedia.org/w/index.php?title=File:Stromatolites.jpg *License:* Public Domain *Contributors:* P. Carrara, NPS ... 6
Figure 3 *Source:* https://en.wikipedia.org/w/index.php?title=File:Darwin_restored2.jpg *License:* Public Domain *Contributors:* Bain News Service, publisher ... 10
Figure 4 *Source:* https://en.wikipedia.org/w/index.php?title=File:Aleksandr_Oparin_and_Andrei_Kursanov_in_enzymology_laboratory_1938.jpg *License:* Public Domain *Contributors:* Andrey Korzun, Butko, Cmapm, Radek Rassel, Salix, 1 anonymous edits ... 11
Figure 5 *Source:* https://en.wikipedia.org/w/index.php?title=File:Phylogenic_Tree-en.svg *License:* Creative Commons Attribution-Sharealike 3.0 *Contributors:* Phylogenic Tree.jpg: John D. Croft derivative work: Conquistador ... 16
Figure 6 *Source:* https://en.wikipedia.org/w/index.php?title=File:Phospholipids_aqueous_solution_structures.svg *License:* Public Domain *Contributors:* Mariana Ruiz Villarreal ,LadyofHats ... 21
Figure 7 *Source:* https://en.wikipedia.org/w/index.php?title=File:10_small_subunit.gif *License:* Public Domain *Contributors:* Animation by David S. Goodsell, RCSB Protein Data Bank ... 23
Figure 8 *Source:* https://en.wikipedia.org/w/index.php?title=File:Blacksmoker_in_Atlantic_Ocean.jpg *License:* Public Domain *Contributors:* P. Rona ... 29
Figure 9 *Source:* https://en.wikipedia.org/w/index.php?title=File:Methane-2D-stereo.svg *License:* Public Domain *Contributors:* Calvero, Daniele Pugliesi, Patricia.fidi, Sarang, Swisstack ... 34
Figure 10 *Source:* https://en.wikipedia.org/w/index.php?title=File:Formation_of_Glycolaldehyde_in_star_dust.png *License:* Public Domain *Contributors:* Lara Clemence ... 36
Figure 11 *Source:* https://en.wikipedia.org/w/index.php?title=File:Buckminsterfullerene-perspective-3D-balls.png *License:* Public Domain *Contributors:* User:Benjah-bmm27 ... 37
Image *Source:* https://en.wikipedia.org/w/index.php?title=File:Lock-green.svg *License:* Creative Commons Zero *Contributors:* User:Trappist the monk ... 47
Image *Source:* https://en.wikipedia.org/w/index.php?title=File:Sound-icon.svg *License:* GNU Lesser General Public License *Contributors:* Crystal SVG icon set ... 49
Image *Source:* https://en.wikipedia.org/w/index.php?title=File:Wiktionary-logo-en-v2.svg *Contributors:* User:Dan Polansky, User:Smurrayinchester 49
Figure 12 *Source:* https://en.wikipedia.org/w/index.php?title=File:Geological_time_spiral.png *License:* Public Domain *Contributors:* United States Geological Survey ... 53
Figure 13 *Source:* https://en.wikipedia.org/w/index.php?title=File:FullMoon2010.jpg *License:* Creative Commons Attribution-Sharealike 3.0 *Contributors:* Gregory H. Revera ... 54
Figure 14 *Source:* https://en.wikipedia.org/w/index.php?title=File:Acasta_gneiss.jpg *License:* Creative Commons Attribution-Sharealike 3.0 *Contributors:* User:Pedroalexandrade ... 55
Figure 15 *Source:* https://en.wikipedia.org/w/index.php?title=File:Cyanobacterial-algal_mat.jpg *License:* Creative Commons Attribution-Sharealike 3.0 *Contributors:* Aleksey Nagovitsyn (User:Alnagov) ... 61
Figure 16 *Source:* https://en.wikipedia.org/w/index.php?title=File:Halobacteria.jpg *License:* Public Domain *Contributors:* NASA ... 61
Figure 17 *Source:* https://en.wikipedia.org/w/index.php?title=File:Endomembrane_system_diagram_en_(edit).svg *License:* Public Domain *Contributors:* Mariana Ruiz Villarreal LadyofHats ... 62
Figure 18 *Source:* https://en.wikipedia.org/w/index.php?title=File:Ceratium_furca.jpg *License:* Creative Commons Attribution-Sharealike 3.0 *Contributors:* Minami Himemiya ... 62
Figure 19 *Source:* https://en.wikipedia.org/w/index.php?title=File:Mikrofoto.de-Blepharisma_japonicum_15.jpg *License:* Creative Commons Attribution-Sharealike 3.0 Germany *Contributors:* Frank Fox ... 63
Figure 20 *Source:* https://en.wikipedia.org/w/index.php?title=File:DickinsoniaCostata.jpg *License:* Creative Commons Attribution 2.5 *Contributors:* Abyssal, Jarould, Kevmin, MGA73bot2, OgreBot 2, Πβ, 3 anonymous edits ... 63
Figure 21 *Source:* https://en.wikipedia.org/w/index.php?title=File:Utatsusaurus_BW.jpg *License:* Creative Commons Attribution 2.5 *Contributors:* Nobu Tamura (http://spinops.blogspot.com) ... 64
Figure 22 *Source:* https://en.wikipedia.org/w/index.php?title=File:Plateosaurus_panorama.jpg *License:* Creative Commons Attribution-Sharealike 3.0 *Contributors:* FunkMonk ... 64
Figure 23 *Source:* https://en.wikipedia.org/w/index.php?title=File:Cycas_circinalis.jpg *License:* GNU Free Documentation License *Contributors:* BotMultichill, MGA73bot2, MPF ... 67
Figure 24 *Source:* https://en.wikipedia.org/w/index.php?title=File:Patriofelis-mount.jpg *License:* Creative Commons Attribution-Sharealike 3.0 *Contributors:* User:Gally242 ... 67
Figure 25 *Source:* https://en.wikipedia.org/w/index.php?title=File:Icaronycteris_index.jpg *License:* Creative Commons Attribution-Sharealike 2.0 *Contributors:* Andrew Savedra ... 68
Figure 26 *Source:* https://en.wikipedia.org/w/index.php?title=File:Grassflowers.jpg *License:* Public Domain *Contributors:* MILEPRI, Nomarcland, OgreBot 2, Thiotrix ... 68
Figure 27 *Source:* https://en.wikipedia.org/w/index.php?title=File:Cms-newyorkzoologicalsociety1910.jpg *License:* Public Domain *Contributors:* New York Zoological Society. ... 69
Figure 28 *Source:* https://en.wikipedia.org/w/index.php?title=File:Lipotes_vexillifer.png *License:* GNU Free Documentation License *Contributors:* Alessio Marrucci ... 69
Figure 29 *Source:* https://en.wikipedia.org/w/index.php?title=File:Diceros_bicornis_longipes.jpg *License:* Public Domain *Contributors:* Senckenberg Museum ... 71
Figure 30 *Source:* https://en.wikipedia.org/w/index.php?title=File:Thylacine_Stuffed_specimen.jpg *License:* Creative Commons Attribution-Sharealike 3.0 *Contributors:* User:Momotarou2012 ... 71
Figure 31 *Source:* https://en.wikipedia.org/w/index.php?title=File:Champagne_vent_white_smokers.jpg *License:* Public Domain *Contributors:* NOAA ... 74
Figure 32 *Source:* https://en.wikipedia.org/w/index.php?title=File:Halobacteria.jpg *License:* Public Domain *Contributors:* NASA ... 74
Figure 33 *Source:* https://en.wikipedia.org/w/index.php?title=File:Porous_chondriteIDP.jpg *License:* Creative Commons Attribution *Contributors:* Basilicofresco, Common Good, Daniele Pugliesi, FSII, File Upload Bot (Magnus Manske), Karelj, Loadmaster, OgreBot 2, Pulsar51, Ruslik0, 1 anonymous edits ... 75
Figure 34 *Source:* https://en.wikipedia.org/w/index.php?title=File:Stromatolites_in_Sharkbay.jpg *License:* GNU Free Documentation License *Contributors:* Paul Harrison ... 76
Figure 35 *Source:* https://en.wikipedia.org/w/index.php?title=File:Stromatolites.jpg *License:* Public Domain *Contributors:* P. Carrara, NPS ... 77
Figure 36 *Source:* https://en.wikipedia.org/w/index.php?title=File:Cyanobacterial-algal_mat.jpg *License:* Creative Commons Attribution-Sharealike 3.0 *Contributors:* Aleksey Nagovitsyn (User:Alnagov) ... 77
Figure 37 *Source:* https://en.wikipedia.org/w/index.php?title=File:Runzelmarken.jpg *License:* Creative Commons Attribution-Sharealike 3.0 *Contributors:* Smith609 ... 77
Figure 38 *Source:* https://en.wikipedia.org/w/index.php?title=File:Kinneyia_Grimsby_Silurian_Niagara_Gorge.jpg *License:* Creative Commons Zero *Contributors:* User:Wilson44691 ... 78
Figure 39 *Source:* https://en.wikipedia.org/w/index.php?title=File:Pasteur's_experiment_testing_spontaneous_generation_and_biogenesis.gif *Contributors:* User:Thebiologyprimer ... 82
Figure 40 *Source:* https://en.wikipedia.org/w/index.php?title=File:Aristotle's_model_of_Inheritance.png *Contributors:* User:Chiswick Chap ... 84
Figure 41 *Source:* https://en.wikipedia.org/w/index.php?title=File:Argopecten_irradians.jpg *License:* Public domain *Contributors:* Rachael Norris and Marina Freudzon / Mayscallop at en.wikipedia ... 86
Figure 42 *Source:* https://en.wikipedia.org/w/index.php?title=File:Pollicipes_cornucopia.jpg *Contributors:* - ... 88
Figure 43 *Source:* https://en.wikipedia.org/w/index.php?title=File:Branta_leucopsis.jpg *License:* Creative Commons Attribution-Sharealike 2.5 *Contributors:* Linnea Samila ... 88

License

Index

www.ingramcontent.com/pod-product-compliance
Lightning Source LLC
Chambersburg PA
CBHW022107210326
41521CB00030B/358